POEMS
PLEAS

BOOK II

POEMS FOR PLEASURE

CHOSEN AND EDITED BY

A. F. SCOTT

BOOK II

CAMBRIDGE UNIVERSITY PRESS

Published by the Syndics of the Cambridge University Press
Bentley House, 200 Euston Road, London, NW1 2DB
American Branch: 32 East 57th Street, New York, N.Y. 10022

ISBN 0 521 06245 4

First published 1955
Reprinted 1955 1959 1961
1963 1967 1972 1974

Printed in Great Britain
at the University Printing House, Cambridge
(Brooke Crutchley, University Printer)

CONTENTS

PART III

§1. THE MAGIC OF WORDS

§2. THE POET'S VISION

§3. NARRATIVE POEMS ON SEA AND LAND

§4. THE POET AND THE MODERN WORLD

PART IV

§1. THE MUSIC OF POETRY

§2. SCENES OF THE MACHINE AGE

§3. STORIES OF PURE IMAGINATION

§4. THE ETERNAL THEME

CONTENTS

INTRODUCTION

It has been said that boys and girls in the secondary school do not like poetry. This perhaps unexpected statement is supported by the evidence of many teachers today. Poetry, they say, cannot compete in attractiveness with music; and in immediacy of appeal painting is far more popular. In the English lesson, prose stories, plays, composition work are all considered to be more interesting and enjoyable than poetry.

This is remarkable, because from the nursery onwards children show a real delight in rhythm and the merry jingle of words. They enjoy nursery rhymes, and like games turning on these rhymes, sing-songing them throughout the house or in the open air. They like phrases that have the touch of poetry about them: namby-pamby, fiddle-faddle, higgledy-piggledy. And the language of poetry so often resembles the language of children, being fresh and imaginative.

Perhaps one reason why boys in the secondary school do not like poetry is that now they are trying to master prose: not only to read it with understanding but to write it. Prose is the medium used for all subjects now being studied—history, geography, science, as well as English. The most important activity, therefore, is to understand the meaning whatever the particular subject may be. Meaning is all important.

It would therefore appear that schoolboys do not like poetry because they cannot always understand it. They look first of all for a meaning, feeling that now they should understand and are exasperated when they do not. When the meaning of a line or of a verse of poetry is carefully explained to them they wonder why the poet could not say whatever he had to say more simply, or even why he did not say it in prose. Some teachers, moreover, stress the importance of meaning too much, so that a poem becomes material for translation from one set of words into another, and the class may well find the translation the more satisfactory!

Yet boys and girls still enjoy the sound of words and the rhythm of phrases, and take an eager delight in music. They are fond of singing and dancing, perhaps even more than when they were younger. They enjoy drawing and painting, and making illustrations for the stories they have read. And they are fascinated by prose stories, by narratives of action. They are now beginning to respond more fully to the emotions they see expressed in films, pantomimes, and plays.

One may, I think, appeal to the capacity to enjoy all these things in presenting poetry. When we consider the simplest elements of poetry we see that a poem consists of a tune, a picture, a story, and a feeling. Though these are closely related, fused together into a single artistic whole, yet it is possible to select poems which will show each particular element above the rest. Poems can be chosen with a very strongly marked rhythm, where sound is more important than the other elements—as in nonsense verse, some early folk-songs, sea shanties, marching songs. Here meaning is of little or no importance. Such poetry may be enjoyed for the sound alone, for the exhilaration the words can give, spoken singly or in chorus, whispered, shouted aloud, moving sometimes as fast as a galloping horse, and sometimes as slowly as a ceremonial procession.

Then a poem can be a picture in words. The young have powers of vivid visualization. Let them read such poems to see trees and birds and animals, the world around them, described with a new clarity and precision. Let them draw what they are now looking at through the eyes of the poet, and so find a new delight in the poet's power to describe. They will enjoy this keener sight, and begin to see things more vividly themselves.

Once schoolboys can appreciate the narrative skill of the old ballads, in close relationship perhaps to the narrative technique of the cinema, they may be moved and excited by the economy, imagination and energy of these tales, and may find ballads and their modern counterparts better than some of the prose narratives they are accustomed to.

Finally, they may be presented with suitable poems expressing the poet's feelings of joy or of unhappiness, and such emotions as anger, grief or pity, which are within their own normal experience and so embodying the important factor of recognition. They may well respond to these expressions of feeling, and later come to appreciate poems revealing the more profound thoughts and emotions.

In his book, *Poetry in School*, Dr J. H. Jagger recalls an eminent lecturer who when instructing his audience of teachers in training said that a poem 'could be studied for the facts it contained, or for its formal characteristics, or for the meaning of the words, or for their derivation, or for its moral (if it contained a perceptible moral), or for its allusions (if there were any), or it could be treated as an historical document (if it had any value of that kind), or it could be correlated with some other subject of instruction, or it could be used as a grammatical exercise, or it could be compared with a standard—given marks, so to speak—or it could be committed to memory'.

Today, our attitude to poetry in schools is mercifully more enlightened, though some of these methods of 'teaching' poetry are still to be found in the class-room. You may notice that though the catalogue above is pretty exhaustive it omits one prime feature (an omission which utterly condemns all the rest) —enjoyment of poetry for its own sake. Wordsworth said that poetry is written to give pleasure.

This selection of poems has been made with this one end constantly in view. They are poems for pleasure, carefully chosen first to recapture the boy's delight in rhythm and picture, in story and feeling, and then leading him progressively to enjoy these four aspects of poetry and to what may be a more complete enjoyment of their summed effect than would otherwise be possible.

The anthology is divided into four parts, one for each school year in the secondary school, and each part is divided into four sections. Introducing each section are brief commentaries on one aspect of poetry under the four main groupings—and each commentary enlarges on what is said in the corresponding commentary in the preceding part. So there is a continuity from one part to the next, and a unity in each individual part.

The anthology covers the first four years in the secondary school. The poems have not only been especially chosen to suit the age and interests of the pupil, but have been carefully arranged by subject so that they bear a close relationship to other poems in the section. This arrangement throws similarities (and sometimes differences) of expression and treatment into strong relief, and gives additional point to the anthology as a whole.

About half the poems are by modern poets, showing that poetry is not something unrelated to modern life, a collection of museum pieces, but a natural and instinctive expression of the human spirit, as much alive and as vital today as in the past.

A. F. SCOTT

June 1954

ACKNOWLEDGMENTS

I should like to acknowledge my deep personal debt to Mr C. E. Carrington for his friendly encouragement and expert advice over a number of years, and especially for his most valued guidance in the compiling of this Anthology. I wish also to express my thanks to the staff of the Cambridge University Press for their patience and continued help.

The author's thanks are due to the following for permission to quote copyright matter:

Mr de la Mare and Messrs Faber & Faber (*The Scarecrow*, *Tartary*, *The Listeners*, *Song of the Mad Prince*, *Nod*); Dr John Masefield, O.M., the Society of Authors and the Macmillan Company, New York (*Cargoes*); the Clarendon Press, Oxford (*London Snow* and *North-wind in October* from *The Shorter Poems of Robert Bridges*); Messrs Macmillan & Co., Ltd. and Mr Ralph Hodgson (*The Bull*, *Time, you old Gipsy Man* and *Eve* from *Poems*); the Hardy Estate and Messrs Macmillan & Co., Ltd. (*Snow in the Suburbs* from *Collected Poems* by Thomas Hardy); Mrs George Bambridge, Messrs Macmillan & Co., Ltd. and the Macmillan Company of Canada (*The Way through the Woods* from *Rewards and Fairies* by Rudyard Kipling); Mrs George Bambridge, Messrs Hodder & Stoughton and the Macmillan Company of Canada (*The Secret of the Machines* from *Rudyard Kipling's Verse: Definitive Edition*); Mrs Edward Thomas and Messrs Faber & Faber (*October*); Mr Robert Graves (*Welsh Incident*, *Outlaws* and *1805* from *Collected Poems 1914–1947* published by Messrs Cassell & Company); Mrs Harold Monro (*Journey* and *Clock*); the Macmillan Company, New York (*Factory Windows* from *Collected Poems* by Vachel Lindsay); Messrs Appleton-Century-

Crofts (*Flower-fed Buffaloes* from *Going to the Stars* by Vachel Lindsay, copyright 1926 D. Appleton & Company, reprinted by permission of the publishers Appleton-Century-Crofts); Mrs W. H. Davies and Messrs Jonathan Cape (*Rich Days* from *The Collected Poems of W. H. Davies*); Mr Robert Frost, Messrs Jonathan Cape and Henry Holt Inc. (*Mowing* and *The Line Gang* from the *Complete Poems of Robert Frost*); Miss D. E. Collins and Messrs Methuen (*Lepanto* from *The Collected Poems of G. K. Chesterton*); Sir John Squire (*The Ship* and *Late Snow*); Messrs Sidgwick & Jackson (*Clouds* from *The Collected Poems of Rupert Brooke*; *Romance* from *Poems* by W. J. Turner; *The Bridge* by J. R. Anderson); Mr W. Gibson and Messrs Macmillan & Co., Ltd. (*The Ice-Cart* and *Flannan Isle*); Harcourt Brace & Company Inc. (*Portrait of a Machine* from *Selected Poems and Parodies of Louis Untermeyer* and *To a Telegraph Pole* from *Burning Bush*); Messrs Gerald Duckworth & Company (*Tarantella* from *Sonnets and Verse* by Hilaire Belloc; *Coromandel* and *Winter the Huntsman* from *Collected Poems* by Osbert Sitwell; and *Rio Grande* from *Selected Poems* by Sacheverell Sitwell); Messrs Faber & Faber (*Prelude I* from *Collected Poems 1909–35* by T.S. Eliot; *Express*, *Landscape near an Aerodrome*, *He will watch the Hawk* and *The Pylons* from *Poems* by Stephen Spender; *Crucifixion of the Skyscraper* from *The Black Rock* by J. Gould Fletcher); Mr W. H. Auden and the G.P.O. (*Night Mail*); Mr Siegfried Sassoon and Messrs Faber & Faber (*Morning Express* and *Everyone Sang*); Messrs Jonathan Cape and the Rev. Dr Andrew Young (*A Windy Day* from *Collected Poems of Andrew Young*); Dr Oliver Gogarty and Messrs Constable (*Time, Gentlemen, Time!*); Mr Richard Church (*The Pigeon*); Dorothy Wellesley (*Docks*); Mr John Davidson and the Richards Press (*A Runnable Stag*); Dame Edith Sitwell and Messrs Macmillan & Co., Ltd. (*Polka* from *The Canticle of the Rose*) and Messrs Blackwell

(*Wooden Pegasus* from *The King of China's Daughter*); Professor V. de S. Pinto (*In the Train*); Messrs Allen & Unwin (*Patch-Shaneen* from *Poems and Translations* by J. M. Synge); Messrs Jonathan Cape and Mr Henry Reed (*The Naming of Parts* from *A Map of Verona*); Professor C. Day Lewis and the Hogarth Press (*Now the full-throated daffodils*); Messrs Little, Brown & Company (*The Chariot* from *The Complete Poems of Emily Dickinson*); Messrs J. M. Dent & Sons (*Poem in October* and *And Death shall have no Dominion* by Dylan Thomas); Lord Gorell and Messrs John Murray (*London to Paris by Air*); Messrs Constable (*To Iron Founders and Others* by Gordon Bottomley); Mr Stanley Snaith (*To some Builders of Cities*); the Hogarth Press (*Beleaguered Cities* by F. L. Lucas); Messrs Heinemann and Mr G. Rostrevor Hamilton (*Tugs*); Messrs John Murray (*Parliament Hill Fields* from *Selected Poems* by John Betjeman); the Estate of the late Stephen Vincent Benet (*Portrait of a Boy*).

A. F. SCOTT

PART I / PART II

[TUNE]

1. Rhythm in Verse

Instinctive enjoyment of dancing. Rhythmical tribal song and dance. From this poetry is born. Rhythm of nursery rhymes. Work songs. Marching songs. Pleasing jingles. Rhythm the first and most important element of poetry.

1. The Poet's Song

Our delight in music. Importance of music in Queen Elizabeth I's time. Popularity of part-songs. Songs at Court and in the theatre. Songs of Shakespeare's plays. Songs then appreciated for own sound. Cut off from music, lyrical poetry is born.

[PICTURE]

2. Pictures in Words

Carefully chosen words effective in description. Examples of Pidgin English. Pictorial and poetic in effect. Definition of a cow. The use of epithets and of verbs. Poetry uses words in a vivid and at same time precise way.

2. The Natural Scene

The use of comparisons in everyday speech. How poetry compares things. Examples of the simile. Examples of the metaphor. The importance of Spring in descriptive poetry. Nature poetry part of our poetic tradition.

[STORY]

3. Tales and Minstrelsy

Our love of stories. Nursery rhymes that tell a story. Origin of nursery rhymes. Folk-songs and traditional tales compared with nursery rhymes. How they were handed down to us. The oral tradition. Different kinds of folk-songs.

3. Ballads Old and New

The way a ballad tells a story compared with the narrative technique of a film. Directness of action. Simplicity and economy of ballad. Quick cutting in the film. No criticism, no probing into motive. Use of details, terse dialogue. Ballad refrain, cf. film music.

[FEELING]

4. The Poet's Feeling

Simple and obvious feelings described. Happiness and unhappiness. How simple feeling when frustrated leads to more complicated state of feeling called emotion. Emotions described. Poet's expression of emotion—pity and grief. Enlarging sympathies.

4. The Poet's Heart

Our earliest feelings are for personal possessions. These later extended to home, friends, school, country. Feelings bound up with idea, such as friendship, called sentiments. Poems expressing sentiments —love of countryside, then of country—patriotism and loyalty.

Index of Authors

Index of First Lines

SCHEME OF BOOK II

PART III

[TUNE]

1. The Magic of Words

Words have not only sound and meaning but associations. Consideration of words 'horse', 'charger', 'steed'; associations of words 'castle', and 'treasure'. The flavour of words—poets' inventions. The magic of sound in a passage full of Proper Names.

[PICTURE]

2. The Poet's Vision

The significance of Spring among primitive people. Myths in which Nature is as real as a living thing. Greek and Roman mythology. Gods and goddesses referred to as though living persons. Personification in poetry. Extension in metaphor. Poets' images revealing a keener vision of the world.

[STORY]

3. Narrative Poems on Sea and Land

Primitive men were hunters and warriors. One of the tribe relates stories of their successful exploits. Beginning of narrative poetry—stories usually founded on fact, dealing with battles on sea and land, hunting and action. Similar subjects treated by poets.

[FEELING]

4. The Poet and the Modern World

World of nature to primitive man peopled with spirits. Tried to establish harmony of human culture with natural environment. All now changed—world of machines. Complete revolution in living—machines hated at first—then accepted, sometimes personified. Have not yet brought the golden age.

Index of Authors

PART IV

[TUNE]

1. The Music of Poetry

The regular rhythm of feet of savages became in time feet of poetic metre. Metre a mechanical pattern. Variety of such patterns—stressed and unstressed syllables. Alliteration, rhyme, assonance; a poem scanned—iambuses, trochees, anapaests, dactyls. Effect of music of poetry.

[PICTURE]

2. Scenes of the Machine Age

The tradition of nature poetry. Now a changed world. New technique in description—experiment and invention. Romance in modern poetry. The adaptation for present poetic needs.

[STORY]

3. Stories of Pure Imagination

Reverie, free association and day-dreaming. Day-dreams shaped and controlled. The importance of the imagination. Romantic poetry. 'The willing suspension of disbelief.' The range of such romantic poetry.

[FEELING]

4. The Eternal Theme

Feelings and their association with values. The highest ideals—supreme values—beauty, truth and goodness. Poetry and these supreme values. Thoughts on life and death. How can we enjoy a sad poem? Poem not mere statement of loss but work of art. Its power to expand in the mind, giving pleasure.

Index of First Lines

PART III

SECTION I

THE MAGIC OF WORDS

Many words have not only meaning and sound but have associations as well. Poets make great use of the associations of words.

Let us consider for a moment the words 'horse', 'steed', and 'charger'. According to the dictionary all three have pretty much the same meaning: 'a solid-hoofed quadruped with a long mane and tail, ridden and used as a beast of burden.' But each of these three words has been used in a different setting and in time gained associations—a close connection of ideas—from that setting.

The word 'steed' makes us think of the remote age of chivalry, of knights in armour with plume and lance, of the legends of the Court of King Arthur, of strange adventures, fights with dragons, and the happy release of captive princesses.

The word 'charger', though still connected with fighting, is less romantic for with it we move from the world of legend to that of reality. It calls to mind a well-mounted officer, leading a cavalry charge in a battle fought not very long ago.

'Horse' is a far more general term, and has widely varied associations. We may think of many different kinds of horse: race-horses, hunters, farm-horses, and may perhaps see a horse ploughing a field, or pulling a milk-cart.

Each of these words, 'horse', 'steed', 'charger', because of the associations which have gathered round it, has its own distinctive flavour.

Words are constantly being enriched by associations. Sieges and sorties, battlements and scaling-ladders are all closely connected with the word 'castle'. Gold and precious stones, mule-trains, the Spanish Main, pirates, old maps, have each added another association to the word 'treasure'.

Poets have a keen sense of the flavour of words, of their sound and associations. They enrich their lines with splendid sounding names of romantic places:

> Manna and dates, in argosy transferr'd
> From Fez; and spiced dainties, everyone
> From silken Samarcand to cedar'd Lebanon.

Sometimes the poet himself invents fine-sounding names for his own delight and ours:

> Three dwarfs there were which lived in an isle,
>   And the name of the isle was Lone,
> And the names of the dwarfs were Alliolyle,
>   Lallerie, Muziomone.

One doesn't need to understand the exact associations of the words in the following passage to catch something of their pageantry, and to be moved by the magic of their sound:

> And what resounds
> In fable or romance of Uther's son,
> Begirt with British and Armoric knights;
> And all who since, baptiz'd or infidel,
> Jousted in Aspramont or Montalban,
> Damasco, or Marocco, or Trebisond;
> Or whom Biserta sent from Afric shore
> When Charlemain with all his peerage fell
> By Fontarabbia.

Such poetry demands to be spoken aloud with the emphasis of oratory, for it carries conviction with accent and alliteration and the inspired repetition of vowel sounds, and should fill the ear like a roll of drums.

Romance

When I was but thirteen or so
  I went into a golden land,
Chimborazo, Cotopaxi
  Took me by the hand.

My father died, my brother too,
  They passed like fleeting dreams.
I stood where Popocatapetl
  In the sunlight gleams.

I dimly heard the Master's voice
  And boys far-off at play,
Chimborazo, Cotopaxi
  Had stolen me away.

I walked in a great golden dream
  To and fro from school—
Shining Popocatapetl
  The dusty streets did rule.

I walked home with a gold dark boy
  And never a word I'd say,
Chimborazo, Cotopaxi
  Had taken my speech away:

I gazed entranced upon his face
  Fairer than any flower—
O shining Popocatapetl
  It was thy magic hour:

The houses, people, traffic seemed
  Thin fading dreams by day,
Chimborazo, Cotopaxi
  They had stolen my soul away!

W. J. TURNER

Tartary

If I were Lord of Tartary,
  Myself and me alone,
My bed should be of ivory,
  Of beaten gold my throne;
And in my court should peacocks flaunt,
And in my forests tigers haunt,
And in my pools great fishes slant
  Their fins athwart the sun.

If I were Lord of Tartary,
  Trumpeters every day
To every meal should summon me,
  And in my courtyard bray;
And in the evenings lamps would shine
Yellow as honey, red as wine,
While harp and flute and mandoline,
  Made music sweet and gay.

If I were Lord of Tartary,
  I'd wear a robe of beads,
White, and gold, and green they'd be—
  And clustered thick as seeds;
And ere should wane the morning-star,
I'd don my robe and scimitar,
And zebras seven should draw my car
  Through Tartary's dark glades.

Lord of the fruits of Tartary,
  Her rivers silver-pale!
Lord of the hills of Tartary,
  Glen, thicket, wood and dale!
Her flashing stars, her scented breeze,
Her trembling lakes, like foamless seas,
Her bird-delighting citron-trees
  In every purple vale!

WALTER DE LA MARE

Eldorado

Gaily bedight,
A gallant knight,
In sunshine and in shadow,
Had journeyed long,
Singing a song,
In search of Eldorado.

But he grew old—
This knight so bold—
And o'er his heart a shadow
Fell as he found
No spot of ground
That looked like Eldorado.

And, as his strength
Failed him at length,
He met a pilgrim shadow:
'Shadow,' said he,
'Where can it be,
This land of Eldorado?'

'Over the mountains
Of the Moon,
Down the valley of the Shadow,
Ride, boldly ride,'
The shade replied,
'If you seek for Eldorado.'

EDGAR ALLAN POE

Portrait of a Boy

After the whipping, he crawled into bed;
Accepting the harsh fact with no great weeping.
How funny uncle's hat had looked striped red!
He chuckled silently. The moon came, sweeping

A black frayed rug of tattered cloud before
In scorning; very pure and pale she seemed,
Flooding his bed with radiance. On the floor
Fat motes danced. He sobbed; closed his eyes and dreamed.

Warm sand flowed round him. Blurts of crimson light
Splashed the white grains like blood. Past the cave's mouth
Shone with a large fierce splendour, wildly bright,
The crooked constellations of the South;
Here the Cross swung; and there, confronting Mars,
The Centaur stormed aside a froth of stars.
Within, great casks like wattled aldermen
Sighed of enormous feasts, and cloth of gold
Glowed on the walls like hot desire. Again
Beside webbed purples from some galleon's hold,
A black chest bore the skull and bones in white
Above a scrawled 'Gunpowder!' By the flames,
Decked out in crimson, gemmed with syenite,
Hailing their fellows by outrageous names
The pirates sat and diced. Their eyes were moons.
'Doubloons!' they said. The words crashed gold.
'Doubloons!'

STEPHEN VINCENT BENET

Time, You Old Gipsy Man

Time, you old gipsy man,
Will you not stay,
Put up your caravan
Just for one day?

All things I'll give you,
Will you be my guest,
Bells for your jennet
Of silver the best,
Goldsmiths shall beat you
A great golden ring,
Peacocks shall bow to you,
Little boys sing,

Oh, and sweet girls will
Festoon you with may,
Time, you old gipsy,
Why hasten away?

Last week in Babylon,
Last night in Rome,
Morning, and in the crush
Under Paul's dome;
Under Paul's dial
You tighten your rein—
Only a moment,
And off once again;
Off to some city
Now blind in the womb,
Off to another
Ere that's in the tomb.

Time, you old gipsy man,
  Will you not stay,
Put up your caravan
  Just for one day?

RALPH HODGSON

Epilogue to 'Hassan'

(AT THE GATE OF THE SUN, BAGDAD, IN OLDEN TIME)

THE MERCHANTS (*together*)

Away, for we are ready to a man!
  Our camels sniff the evening and are glad.
Lead on, O Master of the Caravan,
  Lead on the Merchant-Princes of Bagdad.

THE CHIEF DRAPER

Have we not Indian carpets dark as wine,
  Turbans and sashes, gowns and bows and veils,
And broideries of intricate design,
  And printed hangings in enormous bales?

THE CHIEF GROCER

We have rose-candy, we have spikenard,
Mastic and terebinth and oil and spice,
And such sweet jams meticulously jarred
As God's Own Prophet eats in Paradise.

THE PRINCIPAL JEWS

And we have manuscripts in peacock styles
By Ali of Damascus: we have swords
Engraved with storks and apes and crocodiles,
And heavy beaten necklaces for lords.

THE MASTER OF THE CARAVAN

But you are nothing but a lot of Jews.

PRINCIPAL JEW

Sirs, even dogs have daylight, and we pay.

MASTER OF THE CARAVAN

But who are ye in rags and rotten shoes,
You dirty-bearded, blocking up the way?

THE PILGRIMS

We are the Pilgrims, master; we shall go
Always a little further: it may be
Beyond that last blue mountain barred with snow,
Across that angry or that glimmering sea,

White on a throne or guarded in a cave
There lives a prophet who can understand
Why men were born: but surely we are brave
Who take the Golden Road to Samarkand.

THE CHIEF MERCHANT

We gnaw the nail of hurry. Master, away!

ONE OF THE WOMEN

O turn your eyes to where your children stand.
Is not Bagdad the beautiful? O, stay!

MERCHANTS

We take the Golden Road to Samarkand.

AN OLD MAN

Have you not girls and garlands in your homes,
  Eunuchs and Syrian boys at your command?
Seek not excess: God hateth him who roams!

MERCHANTS

  We take the Golden Road to Samarkand.

HASSAN

Sweet to ride forth at evening from the wells,
  When shadows pass gigantic on the sand,
And softly through the silence beat the bells
  Along the Golden Road to Samarkand.

ISHAK

We travel not for trafficking alone;
  By hotter winds our fiery hearts are fanned:
For lust of knowing what should not be known
  We take the Golden Road to Samarkand.

MASTER OF THE CARAVAN

Open the gate, O watchman of the night!

THE WATCHMAN

  Ho, travellers, I open. For what land
Leave you the dim-moon city of delight?

MERCHANTS (*with a shout*)

  We take the Golden Road to Samarkand!
(*The Caravan passes through the gate*)

THE WATCHMAN (*consoling the women*)

What would ye, ladies? It was ever thus.
  Men are unwise and curiously planned.

A WOMAN

They have their dreams, and do not think of us.

VOICES OF THE CARAVAN (*in the distance singing*)

  We take the Golden Road to Samarkand.

JAMES ELROY FLECKER

Cargoes

Quinquireme of Nineveh from distant Ophir
Rowing home to haven in sunny Palestine,
With a cargo of ivory,
And apes and peacocks,
Sandalwood, cedarwood, and sweet white wine.

Stately Spanish galleon coming from the Isthmus,
Dipping through the Tropics by the palm-green shores,
With a cargo of diamonds,
Emeralds, amethysts,
Topazes, and cinnamon, and gold moidores.

Dirty British coaster with a salt-caked smoke stack
Butting through the Channel in the mad March days,
With a cargo of Tyne coal,
Road-rails, pig-lead,
Firewood, iron-ware, and cheap tin trays.

JOHN MASEFIELD

Lepanto

White founts falling in the Courts of the sun,
And the Soldan of Byzantium is smiling as they run;
There is laughter like the fountains in that face of all men feared,
It stirs the forest darkness, the darkness of his beard,
It curls the blood-red crescent, the crescent of his lips,
For the inmost sea of all the earth is shaken with his ships.
They have dared the white republics up the capes of Italy,
They have dashed the Adriatic round the Lion of the Sea,
And the Pope has cast his arms abroad for agony and loss,
And called the kings of Christendom for swords about the Cross.
The cold Queen of England is looking in the glass;
The shadow of the Valois is yawning at the Mass;
From evening isles fantastical rings faint the Spanish gun,
And the Lord upon the Golden Horn is laughing in the sun.

Dim drums throbbing, in the hills half heard,
Where only on a nameless throne a crownless prince has stirred,
Where, risen from a doubtful seat and half attainted stall,
The last knight of Europe takes weapons from the wall,
The last and lingering troubadour to whom the bird has sung,
That once went singing southward when all the world was young.
In that enormous silence, tiny and unafraid,
Comes up along a winding road the noise of the Crusade.
Strong gongs groaning as the guns boom far,
Don John of Austria is going to the war,
Stiff flags straining in the night-blasts cold,
In the gloom black-purple, in the glint old-gold,
Torchlight crimson on the copper kettle-drums,
Then the tuckets, then the trumpets, then the cannon, and he comes

Don John laughing in the brave beard curled,
Spurning of his stirrups like the thrones of all the world,
Holding his head up for a flag of all the free.
Love-light of Spain—hurrah!
Death-light of Africa!
Don John of Austria
Is riding to the sea.

Mahound is in his paradise above the evening star,
(*Don John of Austria is going to the war.*)
He moves a mighty turban on the timeless houri's knees,
His turban that is woven of the sunsets and the seas.
He shakes the peacock gardens as he rises from his ease,
And he strides among the tree-tops and is taller than the trees,
And his voice through all the garden is a thunder sent to bring
Black Azrael and Ariel and Ammon on the wing,
Giants and the Genii,
Multiplex of wing and eye,
Whose strong obedience broke the sky
When Solomon was king.

They rush in red and purple from the red clouds of the morn,
From temples where the yellow gods shut up their eyes in scorn;
They rise in green robes roaring from the green hells of the sea
Where fallen skies and evil hues and eyeless creatures be;

On them the sea-valves cluster and the grey sea-forests curl,
Splashed with a splendid sickness, the sickness of the pearl;
They swell in sapphire smoke out of the blue cracks of the ground,—
They gather and they wonder and give worship to Mahound.
And he saith, 'Break up the mountains where the hermit-folk can
    hide,
And sift the red and silver sands lest bone of saint abide,
And chase the Giaours flying night and day, not giving rest,
For that which was our trouble comes again out of the west.

We have set the seal of Solomon on all things under sun,
Of knowledge and of sorrow and endurance of things done,
But a noise is in the mountains, in the mountains, and I know
The voice that shook our palaces—four hundred years ago:
It is he that saith not "Kismet"; it is he that knows not Fate;
It is Richard, it is Raymond, it is Godfrey in the gate!
It is he whose loss is laughter when he counts the wager worth,
Put down your feet upon him, that our peace be on the earth.'
For he heard drums groaning and he heard guns jar,
(*Don John of Austria is going to the war.*)
Sudden and still—hurrah!
Bolt from Iberia!
Don John of Austria
Is gone by Alcalar.

St Michael's on his Mountain in the sea-roads of the north,
(*Don John of Austria is girt and going forth.*)
Where the grey seas glitter and the sharp tides shift
And the sea-folk labour and the red sails lift.
He shakes his lance of iron and he claps his wings of stone;
The noise is gone through Normandy; the noise is gone alone;
The North is full of tangled things and texts and aching eyes
And dead is all the innocence of anger and surprise,
And Christian killeth Christian in a narrow dusty room,
And Christian dreadeth Christ that hath a newer face of doom,
And Christian hateth Mary that God kissed in Galilee,
But Don John of Austria is riding to the sea.
Don John calling through the blast and the eclipse,
Crying with the trumpet, with the trumpet of his lips,

Trumpet that sayeth ha!
Domino Gloria!
Don John of Austria
Is shouting to the ships.

King Philip's in his closet with the Fleece about his neck,
(*Don John of Austria is armed upon the deck.*)
The walls are hung with velvet that is black and soft as sin,
And little dwarfs creep out of it and little dwarfs creep in.
He holds a crystal phial that has colours like the moon,
He touches, and it tingles, and he trembles very soon,
And his face is as a fungus of a leprous white and grey
Like plants in the high houses that are shuttered from the day,
And death is in the phial and the end of noble work,
But Don John of Austria has fired upon the Turk.
Don John's hunting, and his hounds have bayed—
Booms away past Italy the rumour of his raid.
Gun upon gun, ha! ha!
Gun upon gun, hurrah!
Don John of Austria
Has loosed the cannonade.

The Pope was in his chapel before day or battle broke,
(*Don John of Austria is hidden in the smoke.*)
The hidden room in man's house where God sits all the year,
The secret window whence the world looks small and very dear.
He sees as in a mirror on the monstrous twilight sea
The crescent of the cruel ships whose name is mystery;
They fling great shadows foe-wards, making Cross and Castle dark,
They veil the plumèd lions on the galleys of St Mark;
And above the ships are palaces of brown, black-bearded chiefs,
And below the ships are prisons, where with multitudinous griefs,
Christian captives sick and sunless, all a labouring race repines
Like a race in sunken cities, like a nation in the mines.
They are lost like slaves that swat, and in the skies of morning hung
The stairways of the tallest gods when tyranny was young.

They are countless, voiceless, hopeless as those fallen or fleeing on
Before the high Kings' horses in the granite of Babylon.

And many a one grows witless in his quiet room in hell
Where a yellow face looks inward through the lattice of his cell,
And he finds his God forgotten, and he seeks no more a sign—
(*But Don John of Austria has burst the battle-line!*)
Don John pounding from the slaughter-painted poop,
Purpling all the ocean like a bloody pirate's sloop,
Scarlet running over on the silvers and the golds,
Breaking of the hatches up and bursting of the holds,
Thronging of the thousands up that labour under sea,
White for bliss and blind for sun and stunned for liberty.
Vivat Hispania!
Domino Gloria!
Don John of Austria
Has set his people free!

Cervantes on his galley sets the sword back in the sheath,
(*Don John of Austria rides homeward with a wreath.*)
And he sees across a weary land a straggling road in Spain,
Up which a lean and foolish knight for ever rides in vain,
And he smiles, but not as Sultans smile, and settles back the blade....
(*But Don John of Austria rides home from the Crusade.*)

G. K. CHESTERTON

Rio Grande

By the Rio Grande
They dance no sarabande
On level banks like lawns above the glassy, lolling tide;
Nor sing they forlorn madrigals
Whose sad note stirs the sleeping gales
Till they wake among the trees, and shake the boughs,
And fright the nightingales;
But they dance in the city, down the public squares,
On the marble pavers with each colour laid in shares,
At the open church doors loud with light within,
At the bell's huge tolling,
By the river music, gurgling, thin,
Through the soft Brazilian air.
The Comendador and the Alguacil are there

On horseback, hid with feathers, loud and shrill
Blowing orders on their trumpets like a bird's sharp bill
Through boughs, like a bitter wind, calling
They shine like steady starlight while those other sparks are falling
In burnished armour, with their plumes of fire,
Tireless, while all others tire.
The noisy streets are empty and hushed is the town
To where, in the square, they dance and the band is playing;
Such a space of silence through the town to the river
That the water murmurs loud
Above the band and crowd together;
And the strains of the sarabande,
More lively than a madrigal,
Go hand in hand
Like the river and its waterfall
As the great Rio Grande rolls down to the sea.
Loud is the marimba's note
Above these half-salt waves,
And louder still the tympanum,
The plectrum and the kettle-drum,
Sullen and menacing
Do these brazen voices ring.
They ride outside,
Above the salt-sea's tide,
Till the ships at anchor there
Hear this enchantment
Of the soft Brazilian air,
By those Southern winds wafted,
Slow and gentle,
Their fierceness tempered
By the air that flows between.

SACHEVERELL SITWELL

On the Coast of Coromandel

On the coast of Coromandel
Dance they to the tunes of Handel;
Chorally, that coral coast
Correlates the bone to ghost,
Till word and limb and note seem one,
Blending, binding act to tone.

All day long they point the sandal
On the coast of Coromandel.
Lemon-yellow legs all bare
Pirouette to peruqued air
From the first green shoots of morn,
Cool as northern hunting-horn,
Till the nightly tropic wind
With its rough-tongued, grating rind
Shatters the frail spires of spice.
Imaged in the lawns of rice
(Mirror-flat and mirror green
Is that lovely water's sheen)
Saraband and rigadoon
Dance they through the purring noon,
While the lacquered waves expand
Golden dragons on the sand—
Dragons that must, steaming, die
From the hot sun's agony—
When elephants, of royal blood,
Plod to bed through lilied mud,
Then evening, sweet as any mango,
Bids them do a gay fandango,
Minuet, jig or gavotte....
How they hate the turkey-trot,
The nautch-dance and the highland fling,
Just as they will never sing
Any music save by Handel
On the coast of Coromandel!

OSBERT SITWELL

Tarantella

Do you remember an Inn,
Miranda?
Do you remember an Inn?
And the tedding and the spreading
Of the straw for a bedding,
And the fleas that tease in the High Pyrenees,
And the wine that tasted of the tar?
And the cheers and the jeers of the young muleteers
(Under the dark of the vine verandah)?
Do you remember an Inn, Miranda,
Do you remember an Inn?
And the cheers and the jeers of the young muleteers
Who hadn't got a penny,
And who weren't paying any,
And the hammer at the doors and the Din?
And the Hip! Hop! Hap!
Of the clap
Of the hands to the twirl and the swirl
Of the girls gone chancing,
Glancing,
Dancing,
Backing and advancing,
Snapping of the clapper to the spin
Out and in—
And the Ting, Tong, Tang of the guitar!
Do you remember an Inn,
Miranda?
Do you remember an Inn?

Never more;
Miranda,
Never more.
Only the high peaks hoar:
And Aragon a torrent at the door.

No sound
In the walls of the Halls where falls

The tread
Of the feet of the dead to the ground.
No sound:
Only the boom
Of the Waterfall like Doom.

HILAIRE BELLOC

Polka

'Tra la la la—
    See me dance the polka,'
Said Mr Wagg like a bear,
'With my top hat
And my whiskers that—
(Tra la la la) trap the Fair.

Where the waves seem chiming haycocks
I dance the polka; there
Stand Venus' children in their gay frocks,—
Maroon and marine,—and stare

To see me fire my pistol
Through the distance blue as my coat;
Like Wellington, Byron, the Marquis of Bristol,
Buzbied great trees float.

While the wheezing hurdy-gurdy
Of the marine wind blows me
To the tune of Annie Rooney, sturdy,
Over the sheafs of sea;

And bright as a seedsman's packet
With zinnias, candytufts chill,
Is Mrs Marigold's jacket
As she gapes at the inn door still,

Where at dawn in the box of the sailor,
Blue as the decks of the sea,
Nelson awoke, crowed like the cocks,
Then back to dust sank he.

And Robinson Crusoe
Rues so
The bright and foxy beer,—
But he finds fresh isles in a negress' smiles,
The poxy doxy dear,

As they watch me dance the polka,'
Said Mr Wagg like a bear,
'In my top hat and my whiskers that—
(Tra la la la) trap the Fair.

Tra la la la la—
Tra la la la la—
Tra la la la la la la la
La
La
La!'

EDITH SITWELL

The King of China's Daughter

The King of China's daughter,
She never would love me
Though I hung my cap and bells upon
Her nutmeg tree.
For oranges and lemons,
The stars in bright blue air,
(I stole them long ago, my dear)
Were dangling there.
The Moon did give me silver pence,
The Sun did give me gold,
And both together softly blew
And made my porridge cold;
But the King of China's daughter
Pretended not to see
When I hung my cap and bells upon
Her nutmeg tree.

THE MAGIC OF WORDS

The King of China's daughter
So beautiful to see
With her face like yellow water, left
Her nutmeg tree.
Her little rope for skipping
She kissed and gave it me—
Made of painted notes of singing-birds
Among the fields of tea.
I skipped across the nutmeg grove,—
I skipped across the sea;
But neither sun nor moon, my dear,
Has yet caught me.

EDITH SITWELL

SECTION 2

THE POET'S VISION

We in England are always glad when Spring comes. It is the return of sunshine, and we begin to think of summer holidays. Our gladness is a pale thing compared with the relief and joy felt by primitive peoples, who found the long winter months tedious and depressing. To them it was a kind of miracle when the dead part of the year began to come to life. And so they created stories about this great event; we find myths in many different countries telling us of the passing of the old year and the rebirth of the new, the death of Winter and the resurrection of Spring.

Nature became as real as a living thing, and when poets wrote about nature they wrote as though spirits lived in the sea and the mountains, in the forests and streams. These spirits became gods and goddesses, nymphs and satyrs, portrayed in statues, in painting, on vases as well as in poetry. The Greeks and Romans worshipping these figures of their natural world, created a mythology which became a treasury of stories for the rest of Europe. So real were the god of the sun, the goddess of the moon, or the nymphs of a fountain that poets referred to them as though they were living persons.

This personification is common in nature poetry, and is part of a great tradition which we to-day may find hard to understand. We can, however, see a picture of a living person when the poet describes Spring in these words:

> O thou with dewy locks who lookest down
> Through the clear windows of the morning,

and when a modern poet says:

> To-day the almond-tree turns pink,
> The first flush of spring;

he also, though less obviously, is personifying Spring.

The poet widened the range of personification to include all kinds of thoughts and ideas and abstract things such as sleep:

> Wrinkled with age, and drenched with dew
> Old Nod, the shepherd, goes.

New words were constantly being created in many interesting ways. The sun was the day's eye, and the name 'daisy' was given to the flower because with its petals like rays it resembled the sun, a delightful metaphor from a personified day. Meaning was often extended from personification to the metaphor, and we say the eye of a needle, the hand of a clock, the foot of a bed; this is again comparison calling us to recognize something which we had not seen or thought of before.

In this and in many other ways the poet gives us images which reveal a wider and keener vision of the world.

To Spring

> O thou with dewy locks, who lookest down
> Through the clear windows of the morning, turn
> Thine angel eyes upon our western isle,
> Which in full choir hails thy approach, O Spring!
>
> The hills tell one another, and the listening
> Valleys hear; all our longing eyes are turn'd
> Up to thy bright pavilions: issue forth
> And let thy holy feet visit our clime!

Come o'er the eastern hills, and let our winds
Kiss thy perfumèd garments; let us taste
Thy morn and evening breath; scatter thy pearls
Upon our lovesick land that mourns for thee.

O deck her forth with thy fair fingers; pour
Thy soft kisses on her bosom; and put
Thy golden crown upon her languish'd head,
Whose modest tresses are bound up for thee.

WILLIAM BLAKE

Now the Full-throated Daffodils

Now the full-throated daffodils,
Our trumpeters in gold,
Call resurrection from the ground
And bid the year be told.

To-day the almond-tree turns pink,
The first flush of spring;
Winds loll and gossip through the town
Her secret whispering.

Now too the bird must try his voice
Upon the morning air;
Down drowsy avenues he cries
A novel great affair.

He tells of royalty to be;
How with her train of rose,
Summer to coronation comes
Through waving wild hedgerows.

To-day crowds quicken in a street,
The fish leaps in the flood:
Look there, gasometer rises,
And here bough swells to bud.

For our love's luck, our stowaway,
Stretches in his cabin;
Our youngster joy barely conceived
Shows up beneath the skin.

Our joy was but a gusty thing
Without sinew or wit,
An infant flyaway; but now
We make a man of it.

C. DAY LEWIS

Nod

Softly along the road of evening,
In a twilight dim with rose,
Wrinkled with age, and drenched with dew
Old Nod, the shepherd, goes.

His drowsy flock streams on before him,
Their fleeces charged with gold,
To where the sun's last beam leans low
On Nod the shepherd's fold.

The hedge is quick and green with briar,
From their sand the conies creep;
And all the birds that fly in heaven
Flock singing home to sleep.

His lambs outnumber a noon's roses,
Yet, when night's shadows fall,
His blind old sheep-dog, Slumber-soon,
Misses not one of all.

His are the quiet steeps of dreamland,
The waters of no-more-pain,
His ram's bell rings 'neath an arch of stars,
'Rest, rest, and rest again.'

WALTER DE LA MARE

The Scarecrow

All winter through I bow my head
  Beneath the driving rain;
The North wind powders me with snow
  And blows me black again;
At midnight under a maze of stars
  I flame with glittering rime,
And stand, above the stubble, stiff
  As mail at morning-prime.
But when that child, called Spring, and all
  His host of children, come,
Scattering their buds and dew upon
  These acres of my home,
Some rapture in my rags awakes;
  I lift void eyes and scan
The skies for crows, those ravening foes
  Of my strange master, Man.
I watch him striding lank behind
  His clashing team, and know
Soon will the wheat swish body high
  Where once lay sterile snow;
Soon shall I gaze across a sea
  Of sun-begotten grain,
Which my unflinching watch hath sealed
  For harvest once again.

WALTER DE LA MARE

Ode to Evening

If aught of oaten stop, or pastoral song,
May hope, chaste Eve, to soothe thy modest ear,
    Like thy own solemn springs,
    Thy springs, and dying gales;

O nymph reserved, while now the bright-hair'd sun
Sits in yon western tent, whose cloudy skirts,
    With brede ethereal wove,
    O'erhang his wavy bed:

Now air is hush'd, save where the weak-ey'd bat
With short shrill shriek flits by on leathern wing,
    Or where the beetle winds
    His small but sullen horn,

As oft he rises 'midst the twilight path,
Against the pilgrim borne in heedless hum:
    Now teach me, maid compos'd,
    To breathe some soften'd strain,

Whose numbers stealing thro' thy darkening vale
May not unseemly with its stillness suit,
    As, musing slow, I hail
    Thy genial lov'd return!

For when thy folding-star arising shows
His paly circlet, at his warning lamp
    The fragrant hours, and elves
    Who slept in buds the day,

And many a nymph who wreathes her brows with sedge,
And sheds the fresh'ning dew, and lovelier still,
    The pensive pleasures sweet
    Prepare thy shadowy car.

Then let me rove some wild and heathy scene,
Or find some ruin 'midst its dreary dells,
    Whose walls more awful nod
    By thy religious gleams.

Or if chill blust'ring winds, or driving rain,
Prevent my willing feet, be mine the hut,
    That from the mountain's side,
    Views wilds, and swelling floods,

And hamlets brown, and dim-discover'd spires,
And hears their simple bell, and marks o'er all
    Thy dewy fingers draw
    The gradual dusky veil.

While Spring shall pour his show'rs, as oft he wont,
And bathe thy breathing tresses, meekest Eve!
    While Summer loves to sport,
    Beneath thy ling'ring light:

While sallow Autumn fills thy lap with leaves,
Or Winter, yelling thro' the troublous air,
    Affrights thy shrinking train,
    And rudely rends thy robes:

So long, regardful of thy quiet rule,
Shall Fancy, Friendship, Science, smiling Peace,
    Thy gentlest influence own,
    And love thy fav'rite name!

WILLIAM COLLINS

Rich Days

Welcome to you, rich Autumn days,
  Ere comes the cold, leaf-picking wind;
When golden stooks are seen in fields,
  All standing arm-in-arm entwined;
And gallons of sweet cider seen
On trees in apples red and green.

With mellow pears that cheat our teeth,
  Which melt that tongues may suck them in,
With cherries red, and blue-black plums,
  Now sweet and soft from stone to skin;
And woodnuts rich, to make us go
Into the loneliest lanes we know.

W. H. DAVIES

To Autumn

Season of mists and mellow fruitfulness!
  Close bosom-friend of the maturing sun;
Conspiring with him how to load and bless
  With fruit the vines that round the thatch-eaves run;
To bend with apples the moss'd cottage-trees,
  And fill all fruit with ripeness to the core;
    To swell the gourd, and plump the hazel shells
With a sweet kernel; to set budding more,
  And still more, later flowers for the bees,
  Until they think warm days will never cease,
    For Summer has o'er-brimm'd their clammy cells.

Who hath not seen thee oft amid thy store?
  Sometimes whoever seeks abroad may find
Thee sitting careless on a granary floor,
  Thy hair soft-lifted by the winnowing wind,
Or on a half-reap'd furrow sound asleep,
  Drows'd with the fume of poppies, while thy hook
    Spares the next swath and all its twinèd flowers;
And sometimes like a gleaner thou dost keep
  Steady thy laden head across a brook;
  Or by a cyder-press, with patient look,
    Thou watchest the last oozings hours by hours.

Where are the songs of Spring? Ay, where are they?
  Think not of them, thou hast thy music too,—
While barrèd clouds bloom the soft-dying day,
  And touch the stubble-plains with rosy hue;
Then in a wailful choir the small gnats mourn
  Among the river sallows, borne aloft
    Or sinking as the light wind lives or dies;
And full-grown lambs loud bleat from hilly bourn;
  Hedge-crickets sing; and now with treble soft
  The red-breast whistles from a garden-croft;
    And gathering swallows twitter in the skies.

JOHN KEATS

Autumn

I love the fitful gust that shakes
  The casement all the day,
And from the glossy elm-tree takes
  The faded leaves away,
Twirling them by the window pane
With thousand others down the lane.

I love to see the shaking twig
  Dance till the shut of eve,
The sparrow on the cottage rig,
  Whose chirp would make believe
That Spring was just now flirting by
In Summer's lap with flowers to lie.

I love to see the cottage smoke
  Curl upwards through the trees,
The pigeons nestled round the cote
  On November days like these;
The cock upon the dunghill crowing,
The mill sails on the heath a-going.

The feather from the raven's breast
  Falls on the stubble lea,
The acorns near the old crow's nest
  Drop pattering down the tree;
The grunting pigs, that wait for all,
Scramble and hurry where they fall.

JOHN CLARE

October

The green elm with the one great bough of gold
Lets leaves into the grass slip, one by one,—
The short hill grass, the mushrooms small, milk-white,
Harebell and scabious and tormentil,
That blackberry and gorse, in dew and sun,
Bow down to; and the wind travels too light
To shake the fallen birch leaves from the fern;

The gossamers wander at their own will.
At heavier steps than birds' the squirrels scold.
The rich scene has grown fresh again and new
As Spring and to the touch is not more cool
Than it is warm to the gaze: and now I might
As happy be as earth is beautiful,
Were I some other or with earth could turn
In alternation of violet and rose,
Harebell and snowdrop, at their season due,
And gorse that has no time not to be gay.
But if this be not happiness,—who knows?
Some day I shall think this a happy day,
And this mood by the name of melancholy
Shall no more blackened and obscurèd be.

EDWARD THOMAS

Poem in October

It was my thirtieth year to heaven
Woke to my hearing from harbour and neighbour wood
And the mussel pooled and the heron
Priested shore
The morning beckon
With water praying and call of seagull and rook
And the knock of sailing boats on the net webbed wall
Myself to set foot
That second
In the still sleeping town and set forth.

My birthday began with the water-
Birds and the birds of the winged trees flying my name
Above the farms and the white horses
And I rose
In rainy autumn
And walked abroad in a shower of all my days.
High tide and the heron dived when I took the road
Over the border
And the gates
Of the town closed as the town awoke.

THE POET'S VISION

A springful of larks in a rolling
Cloud and the roadside bushes brimming with whistling
Blackbirds and the sun of October
Summery
On the hill's shoulder,
Here were fond climates and sweet singers suddenly
Come in the morning where I wandered and listened
To the rain wringing
Wind blow cold
In the wood faraway under me.

Pale rain over the dwindling harbour
And over the sea wet church the size of a snail
With its horns through mist and the castle
Brown as owls
But all the gardens
Of spring and summer were blooming in the tall tales
Beyond the border and under the lark full cloud.
There could I marvel
My birthday
Away but the weather turned around.

It turned away from the blithe country
And down the other air and the blue altered sky
Streamed again a wonder of summer
With apples
Pears and red currants
And I saw in the turning so clearly a child's
Forgotten mornings when he walked with his mother
Through the parables
Of sunlight
And the legends of the green chapels

And the twice told fields of infancy
That his tears burned my cheeks and his heart moved in mine.
These were the woods the river and sea
Where a boy
In the listening
Summertime of the dead whispered the truth of his joy
To the trees and the stones and the fish in the tide.

And the mystery
Sang alive
Still in the water and singingbirds.

And there could I marvel my birthday
Away but the weather turned around. And the true
Joy of the long dead child sang burning
In the sun.
It was my thirtieth
Year to heaven stood there then in the summer noon
Though the town below lay leaved with October blood.
O may my heart's truth
Still be sung
On this high hill in a year's turning.

DYLAN THOMAS

North-wind in October

In the golden glade the chestnuts are fallen all,
From the sered boughs of the oak the acorns fall:
The beech scatters her ruddy fire;
The lime hath stripped to the cold,
And standeth naked above her yellow attire:
The larch thinneth her spire
To lay the ways of the wood with cloth of gold.

Out of the golden-green and white
Of the brake the fir-trees stand upright
In the forest of flame, and wave aloft
To the blue of heaven their blue-green tuftings soft.
But swiftly in shuddering gloom the splendours fail,
As the harrying North-wind beareth
A cloud of skirmishing hail
The grieved woodland to smite:
In a hurricane through the trees he teareth,
Raking the boughs and the leaves rending,
And whistleth to the descending
Blows of his icy flail.

Gold and snow he mixeth in spite,
And whirleth afar; as away on his winnowing flight
He passeth, and all again for awhile is bright.

ROBERT BRIDGES

A Windy Day

This wind brings all dead things to life,
Branches that lash the air like whips
And dead leaves rolling in a hurry
Or peering in a rabbits' bury
Or trying to push down a tree;
Gates that fly open to the wind
And close again behind,
And fields that are a flowing sea
And make the cattle look like ships;
Straws glistening and stiff
Lying on air as on a shelf
And pond that leaps to leave itself;
And feathers too that rise and float,
Each feather changed into a bird,
And line-hung sheets that crack and strain;
Even the sun-greened coat,
That through so many winds has served,
The scarcecrow struggles to put on again.

ANDREW YOUNG

Ode to the West Wind

I

O wild West Wind, thou breath of Autumn's being,
Thou, from whose unseen presence the leaves dead
Are driven, like ghosts from an enchanter fleeing,

Yellow, and black, and pale, and hectic red,
Pestilence-stricken multitudes: O thou,
Who chariotest to their dark wintry bed

The wingèd seeds, where they lie cold and low,
Each like a corpse within its grave, until
Thine azure sister of the Spring shall blow

Her clarion o'er the dreaming earth, and fill
(Driving sweet buds like flocks to feed in air)
With living hues and odours plain and hill:

Wild Spirit, which art moving everywhere;
Destroyer and preserver; hear, O hear!

II

Thou on whose stream, 'mid the steep sky's commotion,
Loose clouds like earth's decaying leaves are shed,
Shook from the tangled boughs of Heaven and Ocean,

Angels of rain and lightning; there are spread
On the blue surface of thine airy surge,
Like the bright hair uplifted from the head

Of some fierce Maenad, even from the dim verge
Of the horizon to the zenith's height
The locks of the approaching storm. Thou dirge

Of the dying year, to which this closing night
Will be the dome of a vast sepulchre,
Vaulted with all thy congregated might

Of vapours, from whose solid atmosphere
Black rain, and fire, and hail will burst; O hear!

III

Thou who didst waken from his summer dreams
The blue Mediterranean, where he lay,
Lulled by the coil of his crystalline streams,

Beside a pumice isle in Baiae's bay,
And saw in sleep old palaces and towers
Quivering within the wave's intenser day,

All overgrown with azure moss and flowers
So sweet, the sense faints picturing them! Thou
For whose path the Atlantic's level powers

Cleave themselves into chasms, while far below
The sea-blooms and the oozy woods which wear
The sapless foliage of the ocean, know

Thy voice, and suddenly grow gray with fear,
And tremble and despoil themselves: O hear!

IV

If I were a dead leaf thou mightest bear;
If I were a swift cloud to fly with thee;
A wave to pant beneath thy power, and share

The impulse of thy strength, only less free
Than thou, O uncontrollable! If even
I were as in my boyhood, and could be

The comrade of thy wanderings over heaven,
As then, when to outstrip thy skiey speed
Scarce seemed a vision; I would ne'er have striven

As thus with thee in prayer in my sore need.
O! lift me as a wave, a leaf, a cloud!
I fall upon the thorns of life! I bleed!

A heavy weight of hours has chained and bowed
One too like thee: tameless, and swift, and proud.

V

Make me thy lyre, even as the forest is:
What if my leaves are falling like its own!
The tumult of thy mighty harmonies

Will take from both a deep, autumnal tone,
Sweet though in sadness. Be thou, Spirit fierce,
My spirit! Be thou me, impetuous one!

Drive my dead thoughts over the universe
Like withered leaves to quicken a new birth!
And, by the incantation of this verse,

Scatter, as from an unextinguished hearth
Ashes and sparks, my words among mankind!
Be through my lips to unawakened earth

The trumpet of a prophecy! O, wind,
If Winter comes, can Spring be far behind?

P. B. SHELLEY

Clouds

Down the blue night the unending columns press
In noiseless tumult, break and wave and flow,
Now tread the far South, or lift rounds of snow
Up to the white moon's hidden loveliness.

Some pause in their grave wandering comradeless,
And turn with profound gesture vague and slow,
As who would pray good for the world, but know
Their benediction empty as they bless.

They say that the Dead die not, but remain
Near to the rich heirs of their grief and mirth.
I think they ride the calm mid-heaven, as these,
In wise majestic melancholy train,
And watch the moon, and the still-raging seas,
And men, coming and going on the earth.

RUPERT BROOKE

The Cloud

I bring fresh showers for the thirsting flowers,
From the seas and the streams;
I bear light shade for the leaves when laid
In their noon-day dreams.

From my wings are shaken the dews that waken
The sweet buds every one,
When rock'd to rest on their mother's breast,
As she dances about the sun.
I wield the flail of the lashing hail,
And whiten the green plains under,
And then again I dissolve it in rain,
And laugh as I pass in thunder.

I sift the snow on the mountains below,
And their great pines groan aghast;
And all the night 'tis my pillow white,
While I sleep in the arms of the blast.
Sublime on the towers of my skiey bowers,
Lightning my pilot sits,
In a cavern under is fettered the thunder,
It struggles and howls at fits;
Over earth and ocean, with gentle motion,
This pilot is guiding me,
Lured by the love of the genii that move
In the depths of the purple sea;
Over the rills, and the crags, and the hills,
Over the lakes and plains,
Wherever he dream, under mountain or stream,
The Spirit he loves remains;
And I all the while bask in heaven's blue smile,
Whilst he is dissolving in rains.

The sanguine sunrise, with his meteor eyes,
And his burning plumes outspread,
Leaps on the back of my sailing rack,
When the morning-star shines dead.
As on the jag of a mountain crag,
Which an earthquake rocks and swings,
An eagle alit one moment may sit,
In the light of its golden wings.
But when sunset may breathe, from the lit sea beneath,
Its ardours of rest and of love,

And the crimson pall of eve may fall
  From the depth of heaven above,
With wings folded I rest, on mine airy nest,
  As still as a brooding dove.

That orbèd maiden, with white fire laden,
  Whom mortals call the moon,
Glides glimmering o'er my fleece-like floor,
  By the midnight breezes strewn;
And wherever the beat of her unseen feet,
  Which only the angels hear,
May have broken the woof of my tent's thin roof,
  The stars peep behind her and peer;
And I laugh to see them whirl and flee,
  Like a swarm of golden bees,
When I widen the rent in my wind-built tent,
  Till the calm rivers, lakes and seas,
Like strips of the sky fallen through me on high,
  Are each paved with the moon and these.

I bind the sun's throne with a burning zone,
  And the moon's with a girdle of pearl;
The volcanoes are dim, and the stars reel and swim,
  When the whirlwinds my banner unfurl.
From cape to cape, with a bridge-like shape,
  Over a torrent sea,
Sunbeam-proof, I hang like a roof,
  The mountains its columns be.
The triumphal arch, through which I march
  With hurricane, fire, and snow,
When the powers of the air are chain'd to my chair,
  Is the million-coloured bow;
The sphere-fire above its soft colours wove,
  While the moist earth was laughing below.

I am the daughter of earth and water,
  And the nursling of the sky;
I pass through the pores of the ocean and shores;
  I change, but I cannot die.

For after the rain, when with never a stain,
  The pavilion of heaven is bare,
And the winds and sunbeams with their convex gleams,
  Build up the blue dome of air,
I silently laugh at my own cenotaph,
  And out of the caverns of rain,
Like a child from the womb, like a ghost from the tomb,
  I arise and unbuild it again.

P. B. SHELLEY

The Ice Cart

Perched on my city office-stool
I watched with envy while a cool
And lucky carter handled ice...
And I was wandering in a trice
Far from the grey and grimy heat
Of that intolerable street
O'er sapphire berg and emerald floe
Beneath the still, cold ruby glow
Of everlasting Polar night,
Bewildered by the queer half-light,
Until I stumbled unawares
Upon a creek where big white bears
Plunged headlong down with flourished heels,
And floundered after shining seals
Through shivering seas of blinding blue.
And, as I watched them, ere I knew,
I'd stripped and I was swimming too
Among the seal-pack, young and hale,
And thrusting on with threshing tail,
With twist and twirl and sudden leap
Through crackling ice and salty deep,
Diving and doubling with my kind,
Until at last we left behind
Those big white blundering bulks of death,
And lay at length with panting breath

Upon a far untravelled floe
Beneath a gentle drift of snow—
Snow drifting gently fine and white
Out of the endless Polar night,
Falling and falling evermore
Upon that far untravelled shore,
Till I was buried fathoms deep
Beneath that cold white drifting sleep—
Sleep drifting deep,
Deep drifting sleep....

The carter cracked a sudden whip:
I clutched my stool with startled grip,
Awakening to the grimy heat
Of that intolerable street.

W. W. GIBSON

The Reverie of Poor Susan

At the corner of Wood Street, when daylight appears,
Hangs a Thrush that sings loud, it has sung for three years:
Poor Susan has passed by the spot, and has heard
In the silence of morning the song of the Bird.

'Tis a note of enchantment; what ails her? She sees
A mountain ascending, a vision of trees;
Bright volumes of vapour through Lothbury glide,
And a river flows on through the vale of Cheapside.

Green pastures she views in the midst of the dale,
Down which she so often has tripped with her pail;
And a single small cottage, a nest like a dove's,
The one only dwelling on earth that she loves.

She looks, and her heart is in heaven: but they fade,
The mist and the river, the hill and the shade:
The stream will not flow, and the hill will not rise,
And the colours have all passed away from her eyes!

WILLIAM WORDSWORTH

Winter the Huntsman

Through his iron glades
Rides Winter the Huntsman,
All colour fades
As his horn is heard sighing.

Far through the forest
His wild hooves crash and thunder
Till many a mighty branch
Is torn asunder.

And the red reynard creeps
To his hole near the river,
The copper leaves fall
And the bare trees shiver.

As night creeps from the ground,
Hides each tree from its brother
And each dying sound
Reveals yet another.

Is it Winter the Huntsman
Who gallops through his iron glades,
Cracking his cruel whip
To the gathering shades?

OSBERT SITWELL

London Snow

When men were all asleep the snow came flying,
In large white flakes falling on the city brown,
Stealthily and perpetually settling and loosely lying,
  Hushing the latest traffic of the drowsy town;
Deadening, muffling, stifling its murmurs failing;
Lazily and incessantly floating down and down:
  Silently sifting and veiling road, roof, and railing;
Hiding difference, making unevenness even,
Into angles and crevices softly drifting and sailing.

All night it fell, and when full inches seven
It lay in the depth of its uncompacted lightness,
The clouds blew off from a high and frosty heaven;
And all woke earlier for the unaccustomed brightness
Of the winter dawning, the strange unheavenly glare:
The eye marvelled—marvelled at the dazzling whiteness;
The ear hearkened to the stillness of the solemn air;
No sound of wheel rumbling nor of foot falling,
And the busy morning cries came thin and spare.
Then boys I heard, as they went to school, calling,
They gathered up the crystal manna to freeze
Their tongues with tasting, their hands with snowballing;
Or rioted in a drift, plunging up to the knees;
Or peering up from under the white-mossed wonder,
'O look at the trees!' they cried, 'O look at the trees!'
With lessened load a few carts creak and blunder,
Following along the white deserted way,
A country company long dispersed asunder:
When now already the sun, in pale display
Standing by Paul's high dome, spread forth below
His sparkling beams, and awoke the stir of the day.
For now doors open, and war is waged with the snow,
And trains of sombre men, past tale of number,
Tread long brown paths, as toward their toil they go:
But even for them awhile no cares encumber
Their minds diverted; the daily word is unspoken,
The daily thoughts of labour and sorrow slumber
At the sight of the beauty that greets them, for the charm they
have broken.

ROBERT BRIDGES

Snow in the Suburbs

Every branch big with it,
Bent every twig with it;
Every fork like a white web-foot;
Every street and pavement mute:
Some flakes have lost their way, and grope back upward, when
Meeting those meandering down they turn and descend again.

The palings are glued together like a wall,
And there is no waft of wind with the fleecy fall.

A sparrow enters the tree,
Whereon immediately
A snow-lump thrice his own slight size
Descends on him and showers his head and eyes.
And overturns him,
And near inurns him,
And lights on a nether twig, when its brush
Starts off a volley of other lodging lumps with a rush.

The steps are a blanched slope,
Up which, with feeble hope,
A black cat comes, wide-eyed and thin;
And we take him in.

THOMAS HARDY

Late Snow

The heavy train through the dim country went rolling, rolling,
Interminably passing misty snow-covered ploughland ridges
That merged in the snowy sky; came turning meadows, fences,
Came gullies and passed, and ice-coloured streams under frozen bridges.

Across the travelling landscape evenly drooped and lifted
The telegraph wires, thick ropes of snow in the windless air;
They drooped and paused and lifted again to unseen summits,
Drawing the eyes and soothing them, often to a drowsy stare.

Singly in the snow the ghosts of trees were softly pencilled,
Fainter and fainter, in the distance fading, into nothingness gliding,
But sometimes a crowd of the intricate silver trees of fairyland
Passed, close and intensely clear, the phantom world hiding.

O untroubled these moving mantled miles of shadowless shadows,
And lovely the film of falling flakes, so wayward and slack;
But I thought of many a mother-bird screening her nestlings,
Sitting silent with wide bright eyes, snow on her back.

J. C. SQUIRE

The Ship

There was no song nor shout of joy,
    Nor beam of moon or sun,
When she came back from the voyage
    Long ago begun;
But twilight on the waters
    Was quiet and grey,
And she glided steady, steady and pensive,
    Over the open bay.

Her sails were brown and ragged,
    And her crew hollow-eyed,
But their silent lips spoke content
    And their shoulders pride;
Though she had no captives on her deck,
    And in her hold
There were no heaps of corn or timber
    Or silks or gold.

J. C. SQUIRE

The Old Ships

I have seen old ships sail like swans asleep
Beyond the village which men still call Tyre,
With leaden age o'ercargoed, dipping deep
For Famagusta and the hidden sun
That rings black Cyprus with a lake of fire;
And all those ships were certainly so old—
Who knows how oft with squat and noisy gun,
Questing brown slaves or Syrian oranges,
The pirate Genoese
Hell-raked them till they rolled
Blood, water, fruit and corpses up the hold.
But now through friendly seas they softly run,
Painted the mid-sea blue or shore-sea green,
Still patterned with the vine and grapes in gold.

But I have seen,
Pointing her shapely shadows from the dawn
And image tumbled on a rose-swept bay,
A drowsy ship of some yet older day;
And, wonder's breath indrawn,
Thought I—who knows—who knows—but in that same
(Fished up beyond Aeaea, patched up new
—Stern painted brighter blue—)
That talkative, bald-headed seaman came
(Twelve patient comrades sweating at the oar)
From Troy's doom-crimson shore,
And with great lies about his wooden horse
Set the crew laughing, and forgot his course.

It was so old a ship—who knows, who knows?
—And yet so beautiful, I watched in vain
To see the mast burst open with a rose,
And the whole deck put on its leaves again.

JAMES ELROY FLECKER

SECTION 3

NARRATIVE POEMS ON SEA AND LAND

Primitive men were hunters and warriors risking their lives for the well-being of the tribe. The most successful hunting expeditions and the most exciting victories over other tribes would be recalled in moments of relaxation round the camp fire. One of the tribe, more imaginative than the others, would enlarge upon these simple stories of hunting and fighting. As his skill grew he would give a more lengthy recital of the feats of some outstanding warrior or tribal chief. He would narrate their struggles against the forces of nature—storms and floods, and how they pleased or angered the gods they feared. His companions would delight in these tales, related in some simple poetic form, and so the story-teller would become an important man in the eyes of the others.

Narrative poetry has been popular for a long time, telling stories which, like those of the early bards, are usually founded on fact. It differs from the ballad because narrative poetry has a greater variety of verse forms and treats a large number of themes.

It is interesting, however, to notice how many narrative poems deal with famous battles on sea and on land. The poet, like the primitive bard, still finds his theme in fighting, at Agincourt and the Baltic, at Hohenlinden and Waterloo. He still writes with verve on hunting:

The stag of warrant, the wily stag,
For twenty miles and five and five,
He ran, and he never was caught alive,
This stag, this runnable stag.

But he writes about all kinds of exciting subjects: floods and shipwrecks, escapes and hurried journeys, wild animals and bandits.

One thing is common to all good narrative poems. The story comes first. The feelings of the poet, however much they may be roused, are subordinated to the demands of the story, and he uses many different devices to make the story vivid, convincing, and memorable.

The Sands of Dee

'O Mary, go and call the cattle home,
And call the cattle home,
And call the cattle home
Across the sands of Dee';
The western wind was wild and dank with foam,
And all alone went she.

The western tide crept up along the sand,
And o'er and o'er the sand,
And round and round the sand,
As far as eye could see.
The rolling mist came down and hid the land:
And never home came she.

'Oh! is it weed, or fish, or floating hair—
A tress of golden hair,
A drownèd maiden's hair
Above the nets at sea?
Was never salmon yet that shone so fair
Among the stakes on Dee.'

They rowed her in across the rolling foam,
The cruel crawling foam,
The cruel hungry foam,
To her grave beside the sea:
But still the boatmen hear her call the cattle home
Across the sands of Dee.

CHARLES KINGSLEY

The High Tide on the Coast of Lincolnshire, 1571

The old mayor climbed the belfry tower,
  The ringers ran by two, by three;
'Pull, if ye never pulled before;
  Good ringers, pull your best,' quoth he.
'Play uppe, play uppe, O Boston bells!
  Ply all your changes, all your swells,
    Play uppe, "The Brides of Enderby!"'

Men say it was a stolen tyde,—
  The Lord that sent it, he knows all;
But in myne ears doth still abide
  The message that the bells let fall:
And there was naught of strange, beside
The flight of mews and peewits pied,
    By millions crouched on the old sea-wall.

I sat and spun within the doore,
  My thread brake off, I raised myne eyes!
The level sun, like ruddy ore,
  Lay sinking in the barren skies;
And dark against day's golden death
She moved where Lindis wandereth,—
My sonne's faire wife, Elizabeth.

'Cusha! Cusha! Cusha!' calling,
Ere the early dews were falling,
Farre away I heard her song.
'Cusha! Cusha!' all along,
Where the reedy Lindis floweth,
    Floweth, floweth,
From the meads where melick groweth
Faintly came her milking song.

'Cusha! Cusha! Cusha!' calling,
'For the dews will soone be falling;
Leave your meadow grasses mellow,
    Mellow, mellow;
Quit your cowslips, cowslips yellow;

Come uppe, Whitefoot, come uppe, Lightfoot,
Quit the stalks of parsley hollow,
    Hollow, hollow;
Come uppe, Jetty, rise and follow,
From the clovers lift your head;
Come uppe, Whitefoot, come uppe, Lightfoot,
Come uppe, Jetty, rise and follow,
Jetty, to the milking-shed.'

If it be long, aye, long ago,
  When I beginne to think howe long,
Againe I hear the Lindis flow,
  Swift as an arrowe, sharpe and strong;
And all the aire it seemeth mee
  Bin full of floating bells (sayth shee),
That ring the tune of Enderby.

Alle fresh the level pasture lay,
  And not a shadowe mote be seene,
Save where, full fyve good miles away,
  The steeple towered from out the greene:
And lo! the great bell farre and wide
Was heard in all the countryside
    That Saturday at eventide.

The swanherds where their sedges are
  Moved on in sunset's golden breath,
The shepherde lads I heard afarre,
  And my sonne's wife, Elizabeth;
Till floating o'er the grassy sea
Came downe that kyndly message free,
The 'Brides of Mavis Enderby.'

Then some looked uppe into the sky,
  And all along where Lindis flows
To where the goodly vessels lie,
  And where the lordly steeple shows.
They sayde, 'And why should this thing be,
What danger lowers by land or sea?
They ring the tune of Enderby!

'For evil news from Mablethorpe,
  Of pyrate galleys warping downe;
For shippes ashore beyond the scorpe,
  They have not spared to wake the towne;
But while the west bin red to see,
And storms be none, and pyrates flee,
Why ring "The Brides of Enderby"?'

I looked without, and lo! my sonne
  Came riding downe with might and main,
He raised a shout as he drew on,
  Till all the welkin' rang again,
'Elizabeth! Elizabeth!'
(A sweeter woman ne'er drew breath
Than my sonne's wife, Elizabeth.)

'The olde sea wall (he cried) is downe,
  The rising tide comes on apace,
And boats adrift in yonder towne
  Go sailing uppe the market-place.'
He shook as one that looks on death:
'God save you, mother!' straight he said;
'Where is my wife, Elizabeth?'

'Good sonne, where Lindis winds away
  With her two bairns I marked her long;
And ere yon bells beganne to play,
  Afar I heard her milking song.'
He looked across the grassy sea,
To right, to left, 'Ho Enderby!'
They rang 'The Brides of Enderby!'

With that he cried and beat his breast;
  For lo! along the river's bed
A mighty eygre reared his crest,
  And uppe the Lindis raging sped.
It swept with thunderous noises loud;
Shaped like a curling snow-white cloud,
Or like a demon in a shroud.

And rearing Lindis backward pressed,
  Shook all her trembling bankes amaine;
Then madly at the eygre's breast
  Flung uppe her weltering walls again.
Then bankes came down with ruin and rout,—
Then beaten foam flew round about,—
Then all the mighty floods were out.

So farre, so fast the eygre drave,
  The heart had hardly time to beat,
Before a shallow, seething wave
  Sobbed in the grasses at our feet:
The feet had hardly time to flee
Before it brake against the knee,
And all the world was in the sea.

Upon the roofe we sate that night,
  The noise of bells went sweeping by:
I marked the lofty beacon light
  Stream from the church-tower, red and high,—
A lurid mark and dread to see;
And awesome bells they were to me,
That in the dark rang 'Enderby'.

They rang the sailor lads to guide
  From roofe to roofe who fearless rowed;
And I,—my sonne was at my side,
  And yet the ruddy beacon glowed:
And yet he moaned beneath his breath,
'O come in life, or come in death!
O lost! my love, Elizabeth.'

And didst thou visit him no more?
  Thou didst, thou didst, my daughter deare!
The waters laid thee at his doore,
  Ere yet the early dawn was clear.
Thy pretty bairns in fast embrace,
The lifted sun shone on thy face,
Downe drifted to thy dwelling-place.

That flow strewed wrecks about the grass;
  That ebbe swept out the flocks to sea;
A fatal ebbe and flow, alas!
  To manye more than myne and mee:
But each will mourn his own (shee sayeth).
And sweeter woman ne'er drew breath
Than my sonne's wife, Elizabeth.

I shall never hear her more
By the reedy Lindis' shore,
'Cusha, Cusha, Cusha!' calling,
Ere the early dews be falling;
I shall never hear her song,
'Cusha, Cusha!' all along.
Where the sunny Lindis floweth,
    Floweth, floweth;
From the meads where melick groweth,
When the water winding down,
Onward floweth to the town.

I shall never see her more
Where the reeds and rushes quiver,
    Shiver, quiver:
Stand beside the sobbing river,
Sobbing, throbbing, in its falling,
To the sandy lonesome shore,
I shall never hear her calling,
'Leave your meadow grasses mellow,
    Mellow, mellow;
Quit your cowslips, cowslips yellow;
Come uppe, Whitefoot, come uppe, Lightfoot;
Quit your pipes of parsley hollow,
    Hollow, hollow:
Come uppe, Lightfoot, rise and follow;
    Lightfoot, Whitefoot,
From your clovers lift the head;
Come uppe, Jetty, follow, follow,
Jetty, to the milking-shed.'

JEAN INGELOW

The Spanish Armada

Attend all ye who list to hear our noble England's praise,
I tell of the thrice famous deeds she wrought in ancient days,
When that great fleet invincible against her bore in vain
The richest spoils of Mexico, the stoutest hearts of Spain.

It was about the lovely close of a warm summer day,
There came a gallant merchant-ship full sail to Plymouth Bay;
Her crew hath seen Castile's black fleet beyond Aurigny's isle,
At earliest twilight, on the waves lie heaving many a mile;
At sunrise she escaped their van, by God's especial grace;
And the tall Pinta, till the noon, had held her close in chase.
Forthwith a guard at every gun was placed along the wall;
The beacon blazed upon the roof of Edgecumbe's lofty hall;
Many a light fishing-bark put out to pry along the coast;
And with loose rein and bloody spur rode inland many a post.
With his white hair unbonnet'd the stout old sheriff comes;
Behind him march the halberdiers, before him sound the drums;
His yeomen, round the market-cross, make clear an ample space,
For there behoves him to set up the standard of her Grace.
And haughtily the trumpets peal, and gaily dance the bells,
As slow upon the labouring wind the royal blazon swells.
Look how the Lion of the sea lifts up his ancient crown,
And underneath his deadly paw treads the gay Lilies down.
So stalk'd he when he turn'd to flight on that famed Picard field
Bohemia's plume, and Genoa's bow, and Cæsar's eagle shield:
So glared he when at Agincourt in wrath he turn'd to bay
And crush'd and torn beneath his claws the princely hunters lay.
Ho! strike the flagstaff deep, Sir Knight; ho! scatter flowers, fair
maids:
Ho! gunners, fire a loud salute: ho! gallants, draw your blades;
Thou sun, shine on her joyously—ye breezes, waft her wide;
Our glorious SEMPER EADEM—the banner of our pride.
The freshening breeze of eve unfurl'd that banner's massive fold,
The parting gleam of sunshine kiss'd that haughty scroll of gold;
Night sank upon the dusky beach, and on the purple sea,—
Such night in England ne'er had been, nor e'er again shall be!

From Eddystone to Berwick bounds, from Lynn to Milford Bay,
That time of slumber was as bright and busy as the day;
For swift to east and swift to west the ghastly war-flame spread;
High on Saint Michael's Mount it shone—it shone on Beachy Head.
Far on the deep the Spaniard saw, along each southern shire,
Cape beyond cape, in endless range, those twinkling points of fire;
The fisher left his skiff to rock on Tamar's glittering waves,
The rugged miners pour'd to war from Mendip's sunless caves.
O'er Longleat's towers, o'er Cranbourne's oaks, the fiery herald flew;
He roused the shepherds of Stonehenge, the rangers of Beaulieu.
Right sharp and quick the bells all night rang out from Bristol town,
And ere the day three hundred horse had met on Clifton down;
The sentinel on Whitehall-gate look'd forth into the night,
And saw o'erhanging Richmond Hill, the streak of blood-red light.
Then bugle's note and cannon's roar the death-like silence broke,
And with one start, and with one cry, the royal city woke.
At once on all her stately gates arose the answering fires;
At once the wild alarum clash'd from all her reeling spires;
From all the batteries of the Tower peal'd loud the voice of fear;
And all the thousand masts of Thames sent back a louder cheer:
And from the farthest wards was heard the rush of hurrying feet,
And the broad streams of pikes and flags rush'd down each roaring
street:
And broader still became the blaze, and louder still the din,
As fast from every village round the horse came spurring in:
And eastward straight, from wild Blackheath, the warlike errand
went,
And roused in many an ancient hall the gallant squires of Kent.
Southward, from Surrey's pleasant hills flew those bright couriers
forth;
High on bleak Hampstead's swarthy moor they started for the North;
And on, and on, without a pause, untired they bounded still,
All night from tower to tower they sprang; they sprang from hill to
hill:
Till the proud Peak unfurl'd the flag o'er Darwin's rocky dales,
Till like volcanoes flared to Heaven the stormy hills of Wales;
Till twelve fair counties saw the blaze on Malvern's lonely height,
Till stream'd in crimson on the wind the Wrekin's crest of light,

Till broad and fierce the star came forth on Ely's stately fane,
And tower and hamlet rose in arms o'er all the boundless plain;
Till Belvoir's lordly terraces the sign to Lincoln sent,
And Lincoln sped the message on o'er the wide vale of Trent;
Till Skiddaw saw the fire that burn'd on Gaunt's embattled pile,
And the red glare on Skiddaw roused the burghers of Carlisle.

LORD MACAULAY

The 'Revenge'

(A BALLAD OF THE FLEET)

I

At Flores in the Azores Sir Richard Grenville lay,
And a pinnace, like a flutter'd bird, came flying from far away:
'Spanish ships of war at sea! we have sighted fifty-three!'
Then sware Lord Thomas Howard: ''Fore God I am no coward;
But I cannot meet them here, for my ships are out of gear,
And the half my men are sick. I must fly, but follow quick.
We are six ships of the line; can we fight with fifty-three?'

II

Then spake Sir Richard Grenville: 'I know you are no coward;
You fly them for a moment to fight with them again.
But I've ninety men and more that are lying sick ashore.
I should count myself the coward if I left them, my Lord Howard,
To these Inquisition dogs and the devildoms of Spain.'

III

So Lord Howard past away with five ships of war that day,
Till he melted like a cloud in the silent summer heaven;
But Sir Richard bore in hand all his sick men from the land
Very carefully and slow,
Men of Bideford in Devon,
And we laid them on the ballast down below;
For we brought them all aboard.
And they blest him in their pain, that they were not left to Spain,
To the thumbscrew and the stake, for the glory of the Lord.

IV

He had only a hundred seamen to work the ship and to fight,
And he sailed away from Flores till the Spaniard came in sight,
With his huge sea-castles heaving upon the weather bow.
'Shall we fight or shall we fly?
Good Sir Richard, tell us now,
For to fight is but to die!
There'll be little of us left by the time this sun be set.'
And Sir Richard said again: 'We be all good English men.
Let us bang these dogs of Seville, the children of the devil,
For I never turn'd my back upon Don or devil yet.'

V

Sir Richard spoke and he laugh'd, and we roar'd a hurrah, and so
The little *Revenge* ran on sheer into the heart of the foe,
With her hundred fighters on deck, and her ninety sick below;
For half of their fleet to the right and half to the left were seen,
And the little *Revenge* ran on thro' the long sea-lane between.

VI

Thousands of their soldiers look'd down from their decks and laugh'd,
Thousands of their seamen made mock at the mad little craft
Running on and on, till delay'd
By their mountain-like *San Philip* that, of fifteen hundred tons,
And up-shadowing high above us with her yawning tiers of guns,
Took the breath from our sails, and we stay'd.

VII

And while now the great *San Philip* hung above us like a cloud
Whence the thunderbolt will fall
Long and loud,
Four galleons drew away
From the Spanish fleet that day,
And two upon the larboard and two upon the starboard lay,
And the battle-thunder broke from them all.

VIII

But anon the great *San Philip*, she bethought herself and went
Having that within her womb that had left her ill content;

And the rest they came aboard us, and they fought us hand to hand,
For a dozen times they came with their pikes and musqueteers,
And a dozen times we shook 'em off as a dog that shakes his ears
When he leaps from the water to the land.

IX

And the sun went down, and the stars came out far over the summer sea,
But never a moment ceased the fight of the one and the fifty-three.
Ship after ship, the whole night long, their high-built galleons came,
Ship after ship, the whole night long, with her battle-thunder and flame;
Ship after ship, the whole night long, drew back with her dead and her shame.
For some were sunk and many were shatter'd, and so could fight us no more—
God of battles, was ever a battle like this in the world before?

X

For he said 'Fight on! fight on!'
Tho' his vessel was all but a wreck;
And it chanced that, when half of the short summer night was gone,
With a grisly wound to be drest he had left the deck,
But a bullet struck him that was dressing it suddenly dead,
And himself he was wounded again in the side and the head,
And he said 'Fight on! fight on!'

XI

And the night went down, and the sun smiled out far over the summer sea,
And the Spanish fleet with broken sides lay round us all in a ring;
But they dared not touch us again, for they fear'd that we still could sting,
So they watch'd what the end would be.
And we had not fought them in vain,
But in perilous plight were we,
Seeing forty of our poor hundred were slain,
And half of the rest of us maim'd for life
In the crash of the cannonades and the desperate strife;

And the sick men down in the hold were most of them stark and
cold,
And the pikes were all broken or bent, and the powder was all of it
spent;
And the masts and the rigging were lying over the side;
But Sir Richard cried in his English pride,
'We have fought such a fight for a day and a night
As may never be fought again!
We have won great glory, my men!
And a day less or more
At sea or shore,
We die—does it matter when?
Sink me the ship, Master Gunner—sink her, split her in twain!
Fall into the hands of God, not into the hands of Spain!'

XII

And the gunner said 'Ay, ay,' but the seamen made reply:
'We have children, we have wives,
And the Lord hath spared our lives.
We will make the Spaniard promise, if we yield, to let us go;
We shall live to fight again and to strike another blow.'
And the lion there lay dying, and they yielded to the foe.

XIII

And the stately Spanish men to their flagship bore him then,
Where they laid him by the mast, old Sir Richard caught at last,
And they praised him to his face with their courtly foreign grace;
But he rose upon their decks, and he cried:
'I have fought for Queen and Faith like a valiant man and true;
I have only done my duty as a man is bound to do:
With a joyful spirit I Sir Richard Grenville die!'
And he fell upon their decks, and he died.

XIV

And they stared at the dead that had been so valiant and true,
And had holden the power and glory of Spain so cheap
That he dared her with one little ship and his English few;
Was he devil or man? He was devil for aught they knew,
But they sank his body with honour down into the deep,

And they mann'd the *Revenge* with a swarthier alien crew,
And away she sail'd with her loss and long'd for her own;
When a wind from the lands they had ruin'd awoke from sleep,
And the water began to heave and the weather to moan,
And or ever that evening ended a great gale blew,
And a wave like the wave that is raised by an earthquake grew,
Till it smote on their hulls and their sails and their masts and their
flags,
And the whole sea plunged and fell on the shot-shatter'd navy of
Spain,
And the little *Revenge* herself went down by the island crags
To be lost evermore in the main.

LORD TENNYSON

The Loss of the 'Royal George'

Toll for the brave—
The brave! that are no more:
All sunk beneath the wave,
Fast by their native shore.
Eight hundred of the brave,
Whose courage well was tried,
Had made the vessel heel
And laid her on her side;
A land-breeze shook the shrouds,
And she was overset;
Down went the *Royal George*
With all her crew complete.

Toll for the brave—
Brave Kempenfelt is gone,
His last sea-fight is fought,
His work of glory done.
It was not in the battle,
No tempest gave the shock,
She sprang no fatal leak,
She ran upon no rock;

His sword was in the sheath,
His fingers held the pen,
When Kempenfelt went down
With twice four hundred men.

Weigh the vessel up,
Once dreaded by our foes,
And mingle with your cup
The tears that England owes;
Her timbers yet are sound,
And she may float again,
Full charged with England's thunder,
And plough the distant main;
But Kempenfelt is gone,
His victories are o'er;
And he and his eight hundred
Must plough the wave no more.

WILLIAM COWPER

The Yarn of the 'Nancy Bell'

'Twas on the shores that round our coast
From Deal to Ramsgate span,
That I found alone on a piece of stone
An elderly naval man.

His hair was weedy, his beard was long,
And weedy and long was he,
And I heard this wight on the shore recite,
In a singular minor key:

'Oh, I am a cook and a captain bold,
And the mate of the *Nancy* brig,
And a bo'sun tight, and a midshipmite,
And the crew of the captain's gig.'

And he shook his fists and he tore his hair,
Till I really felt afraid,
For I couldn't help thinking the man had been drinking,
And so I simply said:

'Oh, elderly man, it's little I know
  Of the duties of men of the sea,
And I'll eat my hand if I understand
  How you can possibly be

At once a cook, and a captain bold,
  And the mate of the *Nancy* brig,
And a bo'sun tight, and a midshipmite,
  And the crew of the captain's gig.'

Then he gave a hitch to his trousers, which
  Is a trick all seamen larn,
And having got rid of a thumping quid,
  He spun his painful yarn:

"'Twas in the good ship *Nancy Bell*
  That we sailed to the Indian Sea,
And there on a reef we come to grief,
  Which has often occurred to me.

And pretty nigh all the crew was drowned
  (There was seventy-seven o' soul),
And only ten of the *Nancy's* men
  Said "here" to the muster-roll.

There was me and the cook and the captain bold,
  And the mate of the *Nancy* brig,
And the bo'sun tight, and a midshipmite,
  And the crew of the captain's gig.

For a month we'd neither wittles nor drink,
  Till a-hungry we did feel,
So we drawed a lot, and accordin' shot
  The captain for our meal.

The next lot fell to the *Nancy's* mate,
  And a delicate dish he made;
Then our appetite with the midshipmite
  We seven survivors stayed.

And then we murdered the bos'un tight,
  And he much resembled pig;
Then we wittled free, did the cook and me,
  On the crew of the captain's gig.

Then only the cook and me was left,
  And the delicate question, "Which
Of us two goes to the kettle?" arose,
  And we argued it out as sich.

For I loved that cook as a brother, I did,
  And the cook he worshipped me;
But we'd both be blowed if we'd either be stowed
  In the other chap's hold, you see.

"I'll be eat if you dines off me," says Tom.
  "Yes, that," says I, "you'll be,—
I'm boiled if I die, my friend," quoth I.
  And "Exactly so," quoth he.

Says he, "Dear James, to murder me
  Were a foolish thing to do,
For don't you see that you can't cook *me*,
  While I can—and will—cook *you*!"

So he boils the water, and takes the salt
  And the pepper in portions true
(Which he never forgot), and some chopped shallot,
  And some sage and parsley too.

"Come here," says he, with a proper pride,
  Which his smiling features tell,
"'Twill soothing be if I let you see
  How extremely nice you'll smell."

And he stirred it round and round and round,
  And he sniffed at the foaming froth;
When I ups with his heels, and smothers his squeals
  In the scum of the boiling broth.

And I eat that cook in a week or less,
  And—as I eating be
The last of his chops, why, I almost drops,
  For a vessel in sight I see.

* * *

And I never larf, and I never smile,
  And I never lark or play,
But sit and croak, and a single joke
  I have,—which is to say:

Oh, I am a cook and a captain bold,
  And the mate of the *Nancy* brig,
And a bos'un tight, and a midshipmite,
  And the crew of the captain's gig.'

W. S. GILBERT

The War Song of Dinas Vawr

The mountain sheep are sweeter,
But the valley sheep are fatter;
We therefore deemed it meeter
To carry off the latter.
We made an expedition;
We met a host and quelled it;
We forced a strong position,
And killed the men who held it.

On Dyfed's richest valley,
Where herds of kine were browsing,
We made a mighty sally,
To furnish our carousing.
Fierce warriors rushed to meet us;
We met them, and o'erthrew them:
They struggled hard to beat us;
But we conquered them, and slew them.

As we drove our prize at leisure,
The king marched forth to catch us:
His rage surpassed all measure,
But his people could not match us.
He fled to his hall-pillars;
And, ere our force we led off,
Some sacked his house and cellars,
While others cut his head off.

We there, in strife bewildering,
Spilt blood enough to swim in:
We orphaned many children,
And widowed many women.
The eagles and the ravens
We glutted with our foemen;
The heroes and the cravens,
The spearmen and the bowmen.

We brought away from battle,
And much their land bemoaned them,
Two thousand head of cattle,
And the head of him who owned them:
Ednyfed, King of Dyfed,
His head was borne before us;
His wine and beasts supplied our feasts,
And his overthrow, our chorus.

T. L. PEACOCK

The Destruction of Sennacherib

The Assyrian came down like the wolf on the fold,
And his cohorts were gleaming in purple and gold;
And the sheen of their spears was like stars on the sea,
When the blue wave rolls nightly on deep Galilee.

Like the leaves of the forest when summer is green,
That host with their banners at sunset were seen:
Like the leaves of the forest when autumn hath blown,
That host on the morrow lay withered and strown.

For the Angel of Death spread his wings on the blast,
And breathed on the face of the foe as he passed:
And the eyes of the sleepers waxed deadly and chill,
And their hearts but once heaved, and for ever grew still!

And there lay the steed with his nostril all wide,
But through it there rolled not the breath of his pride;
And the foam of his gasping lay white on the turf,
And cold as the spray of the rock-beating surf.

And there lay the rider, distorted and pale,
With the dew on his brow and the rust on his mail;
The tents were all silent, the banners alone,
The lances unlifted, the trumpet unblown.

And the widows of Ashur are loud in their wail,
And the idols are broke in the temple of Baal;
And the might of the Gentile, unsmote by the sword,
Hath melted like snow in the glance of the Lord!

LORD BYRON

The Ballad of Agincourt

Fair stood the wind for France,
When we our sails advance,
Nor now to prove our chance,
  Longer will tarry;
But putting to the main,
At Caux, the mouth of Seine,
With all his martial train,
  Landed King Harry.

And taking many a fort,
Furnished in warlike sort,
Marcheth tow'rds Agincourt
  In happy hour;
Skirmishing day by day
With those that stopped his way,
Where the French general lay,
  With all his power.

Which in his height of pride,
King Henry to deride,
His ransom to provide
To the king sending.
Which he neglects the while,
As from a nation vile,
Yet with an angry smile,
Their fall portending.

And turning to his men,
Quoth our brave Henry then,
Though they to one be ten,
Be not amazèd.
Yet have we well begun,
Battles so bravely won,
Have ever to the sun
By fame been raisèd.

And, for myself (quoth he),
This my full rest shall be,
England ne'er mourn for me,
Nor more esteem me.
Victor I will remain,
Or on this earth lie slain,
Never shall she sustain
Loss to redeem me.

Poitiers and Cressy tell,
When most their pride did swell,
Under our swords they fell,
No less our skill is,
Than when our grandsire great,
Claiming the regal seat,
By many a warlike feat,
Lopped the French lilies.

The Duke of York so dread,
The eager vaward led;
With the main, Henry sped,
Amongst his henchmen.

Excester had the rear,
A braver man not there;
O Lord, how hot they were
    On the false Frenchmen!

They now to fight are gone,
Armour on armour shone,
Drum now to drum did groan,
    To hear was wonder;
That with the cries they make,
The very earth did shake,
Trumpet to trumpet spake,
    Thunder to thunder.

Well it thine age became,
O noble Erpingham,
Which didst the signal aim
    To our hid forces;
When from a meadow by,
Like a storm suddenly
The English archery
    Stuck the French horses.

With Spanish yew so strong,
Arrows a cloth-yard long,
That like to serpents stung,
    Piercing the weather;
None from his fellow starts,
But playing manly parts,
And like true English hearts
    Stuck close together.

When down their bows they threw,
And forth their bilbos drew.
And on the French they flew,
    Not one was tardy;
Arms were from shoulders sent,
Scalps to the teeth were rent,
Down the French peasants went,
    Our men were hardy.

This while our noble King,
His broadsword brandishing,
Down the French host did ding,
    As to o'erwhelm it;
And many a deep wound lent,
His arms with blood besprent,
And many a cruel dent
    Bruisèd his helmet.

Gloucester, that Duke so good,
Next of the royal blood,
For famous England stood,
    With his brave brother;
Clarence, in steel so bright,
Though but a maiden knight,
Yet in that famous fight,
    Scarce such another.

Warwick in blood did wade,
Oxford the foe invade,
And cruel slaughter made,
    Still as they ran up;
Suffolk his axe did ply,
Beaumont and Willoughby
Bare them right doughtily,
    Ferrers and Fanhope.

Upon Saint Crispin's Day
Fought was this noble fray,
Which fame did not delay
    To England to carry.
O when shall Englishmen
With such acts fill a pen,
Or England breed again
    Such a King Harry?

MICHAEL DRAYTON

The Battle of the Baltic

I

Of Nelson and the North,
Sing the glorious day's renown,
When to battle fierce came forth
All the might of Denmark's crown,
And her arms along the deep proudly shone;
By each gun the lighted brand,
In a bold determined hand,
And the Prince of all the land
Led them on.—

II

Like leviathans afloat,
Lay their bulwarks on the brine,
While the sign of battle flew
On the lofty British line:
It was ten of April morn by the chime:
As they drifted on their path,
There was silence deep as death;
And the boldest held his breath
For a time.

III

But the might of England flush'd
To anticipate the scene;
And her van the fleeter rush'd
O'er the deadly space between.
'Hearts of oak!' our captains cried; when each gun
From its adamantine lips
Spread a death-shade round the ships,
Like the hurricane eclipse
Of the sun.

IV

Again! again! again!
And the havoc did not slack,
Till a feeble cheer the Dane,
To our cheering sent us back;

Their shots along the deep slowly boom:
Then ceased—and all is wail,
As they strike the shatter'd sail,
Or, in conflagration pale,
Light the gloom.

V

Out spoke the victor then,
As he hail'd them o'er the wave:
'Ye are brothers! ye are men!
And we conquer but to save;
So peace instead of death let us bring;
But yield, proud foe, thy fleet,
With the crews, at England's feet,
And make submission meet
To our King.'

VI

Then Denmark bless'd our chief,
That he gave her wounds repose;
And the sounds of joy and grief
From her people wildly rose,
As death withdrew his shades from the day.
While the sun look'd smiling bright
O'er a wide and woeful sight,
Where the fires of funeral light
Died away.

VII

Now joy, old England, raise!
For the tidings of thy might,
By the festal cities' blaze,
While the wine-cup shines in light;
And yet amidst that joy and uproar,
Let us think of them that sleep,
Full many a fathom deep,
By thy wild and stormy steep,
Elsinore!

VIII

Brave hearts! to Britain's pride
Once so faithful and so true,
On the deck of fame that died,
With the gallant good Riou;
Soft sigh the winds of heaven o'er their grave!
While the billow mournful rolls,
And the mermaid's song condoles,
Singing glory to the souls
Of the brave!

THOMAS CAMPBELL

1805

At Viscount Nelson's lavish funeral,
While the mob milled and yelled about St Paul's,
A General chatted with an Admiral:

'One of your Colleagues, Sir, remarked today
That Nelson's *exit*, though to be lamented,
Falls not inopportunely, in its way.'

'He was a thorn in our flesh,' came the reply—
'The most bird-witted, unaccountable,
Odd little runt that ever I did spy.

'One arm, one peeper, vain as Pretty Poll,
A meddler, too, in foreign politics
And gave his heart in pawn to a plain moll.

'He would dare lecture us Sea Lords, and then
Would treat his ratings as though men of honour
And play at leap-frog with his midshipmen!

'We tried to box him down, but up he popped,
And when he'd banged Napoleon at the Nile
Became too much the hero to be dropped.

'You've heard that Copenhagen "blind eye" story?
We'd tied him to Nurse Parker's apron-strings—
By G-d, he snipped them through and snatched the glory!'

'Yet,' cried the General, 'six-and-twenty sail
    Captured or sunk by him off Trafalgar—
That writes a handsome *finis* to the tale.'

'Handsome enough. The seas are England's now.
    That fellow's foibles need no longer plague us.
He died most creditably, I'll allow.'

'And, Sir, the secret of his victories?'
    'By his unServicelike, familiar ways, Sir,
He made the whole Fleet love him, damn his eyes!'

ROBERT GRAVES

Hohenlinden

On Linden, when the sun was low,
All bloodless lay the untrodden snow;
And dark as winter was the flow
    Of Iser, rolling rapidly.

But Linden saw another sight,
When the drum beat at dead of night
Commanding fires of death to light
    The darkness of her scenery.

By torch and trumpet fast array'd
Each horseman drew his battle-blade,
And furious every charger neigh'd
    To join the dreadful revelry.

Then shook the hills with thunder riven;
Then rush'd the steed, to battle driven;
And louder than the bolts of Heaven
    Far flash'd the red artillery.

But redder yet that light shall glow
On Linden's hills of stainèd snow;
And bloodier yet the torrent flow
    Of Iser, rolling rapidly.

'Tis morn; but scarce yon level sun
Can pierce the war-clouds, rolling dun,
Where furious Frank and fiery Hun
Shout in their sulphurous canopy.

The combat deepens. On, ye Brave
Who rush to glory, or the grave!
Wave, Munich, all thy banners wave,
And charge with all thy chivalry!

Few, few shall part, where many meet!
The snow shall be their winding-sheet,
And every turf beneath their feet
Shall be a soldier's sepulchre.

THOMAS CAMPBELL

Waterloo

There was a sound of revelry by night,
And Belgium's capital had gathered then
Her Beauty and her Chivalry, and bright
The lamps shone o'er fair women and brave men.
A thousand hearts beat happily; and when
Music arose with its voluptuous swell,
Soft eyes looked love to eyes which spake again,
And all went merry as a marriage bell;
But hush! hark! a deep sound strikes like a rising knell!

Did ye not hear it?—No; 'twas but the wind,
Or the car rattling o'er the stony street;
On with the dance! let joy be unconfined;
No sleep till morn, when Youth and Pleasure meet
To chase the glowing Hours with flying feet.
But hark!—that heavy sound breaks in once more
As if the clouds its echo would repeat;
And nearer, clearer, deadlier than before!
Arm! arm! it is—it is—the cannon's opening roar!

Within a windowed niche of that high hall
Sate Brunswick's fated chieftain; he did hear
That sound, the first amidst the festival,
And caught its tone with Death's prophetic ear;
And when they smiled because he deemed it near,
His heart more truly knew that peal too well
Which stretched his father on a bloody bier,
And roused the vengeance blood alone could quell.
He rushed into the field, and, foremost fighting, fell.

Ah! then and there was hurrying to and fro,
And gathering tears, and tremblings of distress,
And cheeks all pale, which but an hour ago
Blushed at the praise of their own loveliness;
And there were sudden partings, such as press
The life from out young hearts, and choking sighs
Which ne'er might be repeated: who could guess
If ever more should meet those mutual eyes,
Since upon night so sweet such awful morn could rise!

And there was mounting in hot haste: the steed,
The mustering squadron and the clattering car,
Went pouring forward with impetuous speed,
And swiftly forming in the ranks of war:
And the deep thunder peal on peal afar;
And near, the beat of the alarming drum
Roused up the soldier ere the morning star;
While thronged the citizens with terror dumb,
Or whispering with white lips—'The foe! They come! they
come!'

And wild and high the 'Cameron's gathering' rose!
The war-note of Lochiel, which Albyn's hills
Have heard, and heard, too, have her Saxon foes;
How in the noon of night that pibroch thrills,
Savage and shrill! But with the breath which fills
Their mountain-pipe, so fill the mountaineers
With the fierce native daring which instils
The stirring memory of a thousand years,
And Evan's, Donald's fame rings in each clansman's ears!

And Ardennes waves above them her green leaves,
Dewy with Nature's tear-drops, as they pass,
Grieving, if aught inanimate e'er grieves,
Over the unreturning brave,—alas!
Ere evening to be trodden like the grass
Which now beneath them, but above shall grow
In its next verdure, when this fiery mass
Of living valour, rolling on the foe,
And burning with high hope, shall moulder, cold and low.

Last noon beheld them full of lusty life,
Last eve in Beauty's circle proudly gay,
The midnight brought the signal-sound of strife,
The morn the marshalling in arms—the day
Battle's magnificently stern array!
The thunder-clouds close o'er it, which when rent
The earth is covered thick with other clay,
Which her own clay shall cover, heaped and pent,
Rider and horse,—friend, foe,—in one red burial blent!

LORD BYRON

The Burial of Sir John Moore at Corunna

Not a drum was heard, not a funeral note,
As his corpse to the rampart we hurried;
Not a soldier discharged his farewell shot
O'er the grave where our hero we buried.

We buried him darkly at dead of night,
The sods with our bayonets turning;
By the struggling moonbeam's misty light
And the lantern dimly burning.

No useless coffin enclosed his breast,
Not in sheet or in shroud we wound him;
But he lay like a warrior taking his rest,
With his martial cloak around him.

Few and short were the prayers we said,
  And we spoke not a word of sorrow;
But we steadfastly gazed on the face that was dead,
  And we bitterly thought of the morrow.

We thought, as we hollow'd his narrow bed
  And smoothed down his lonely pillow,
That the foe and the stranger would tread o'er his head.
  And we far away on the billow!

Lightly they'll talk of the spirit that's gone
  And o'er his cold ashes upbraid him,—
But little he'll reck, if they let him sleep on
  In the grave where a Briton has laid him.

But half of our heavy task was done
  When the clock struck the hour for retiring:
And we heard the distant and random gun
  That the foe was sullenly firing.

Slowly and sadly we laid him down,
  From the field of his fame fresh and gory;
We carved not a line, and we raised not a stone,
  But we left him alone with his glory.

CHARLES WOLFE

The Cavalier's Escape

Trample! trample! went the roan,
    Trap! trap! went the grey;
But pad! pad! pad! like a thing that was mad,
    My chestnut broke away,
It was just five miles from Salisbury town,
    And but one hour to day.

Thud! thud! came on the heavy roan,
    Rap! rap! the mettled grey;
But my chestnut mare was of blood so rare,
    That she showed them all the way.
Spur on! spur on!—I doffed my hat,
    And wished them all good-day.

They splashed through miry rut and pool—
Splintered through fence and rail;
But chestnut Kate switched over the gate—
I saw them droop and tail.
To Salisbury town, but a mile of down,
Once over this brook and rail.

Trap! trap! I heard their echoing hoofs,
Past the walls of mossy stone;
The roan flew on at a staggering pace,
But blood is better than bone.
I patted old Kate and gave her the spur,
For I knew it was all my own.

But trample! trample! came their steeds,
And I saw their wolf's eyes burn;
I felt like a royal hart at bay,
And made me ready to turn,
I looked where highest grew the may,
And deepest arched the fern.

I flew at the first knave's sallow throat;
One blow, and he was down.
The second rogue fired twice and missed;
I sliced the villain's crown.
Clove through the rest, and flogged brave Kate,
Fast, fast, to Salisbury town.

Pad! pad! they came on the level sward,
Thud! thud! upon the sand;
With a gleam of swords, and a burning match,
And a shaking of flag and hand:
But one long bound, and I passed the gate,
Safe from the canting band.

WALTER THORNBURY

How They Brought the Good News from Ghent to Aix

I sprang to the stirrup, and Joris, and he;
I galloped, Dirck galloped, we galloped all three;
'Good speed!' cried the watch, as the gate-bolts undrew;
'Speed!' echoed the wall to us galloping through;
Behind shut the postern, the lights sank to rest,
And into the midnight we galloped abreast.

Not a word to each other; we kept the great pace,
Neck by neck, stride by stride, never changing our place;
I turned in my saddle and made its girths tight,
Then shortened each stirrup, and set the pique right,
Rebuckled the cheek-strap, chained slacker the bit,
Nor galloped less steadily Roland a whit.

'Twas moonset at starting; but while we drew near
Lokeren, the cocks crew and twilight dawned clear;
At Boom, a great yellow star came out to see;
At Düffeld, 'twas morning as plain as could be;
And from Mecheln church-steeple we heard the half-chime,
So, Joris broke silence with, 'Yet there is time!'

At Aershot, up leaped of a sudden the sun,
And against him the cattle stood black every one,
To stare thro' the mist at us galloping past,
And I saw my stout galloper Roland at last,
With resolute shoulders, each butting away
The haze, as some bluff river headland its spray:

And his low head and crest, just one sharp ear bent back
For my voice, and the other pricked out on his track;
And one eye's black intelligence,—ever that glance
O'er its white edge at me, his own master, askance!
And the thick heavy spume-flakes which aye and anon
His fierce lips shook upwards in galloping on.

By Hasselt, Dirck groaned; and cried Joris, 'Stay spur!
Your Roos galloped bravely, the fault's not in her,

We'll remember at Aix'—for one heard the quick wheeze
Of her chest, saw the stretched neck and staggering knees,
And sunk tail, and horrible heave of the flank,
As down on her haunches she shuddered and sank.

So, we were left galloping, Joris and I,
Past Looz and past Tongres, no cloud in the sky;
The broad sun above laughed a pitiless laugh,
'Neath our feet broke the brittle bright stubble like chaff;
Till over by Dalhem a dome-spire sprang white,
And 'Gallop', gasped Joris, 'for Aix is in sight!'

'How they'll greet us!'—and all in a moment his roan
Rolled neck and croup over, lay dead as a stone;
And there was my Roland to bear the whole weight
Of the news which alone could save Aix from her fate,
With his nostrils like pits full of blood to the brim,
And with circles of red for his eye-sockets' rim.

Then I cast loose my buffcoat, each holster let fall,
Shook off both my jack-boots, let go belt and all,
Stood up in the stirrup, leaned, patted his ear,
Called my Roland his pet-name, my horse without peer;
Clapped my hands, laughed and sang, any noise, bad or good,
Till at length into Aix Roland galloped and stood.

And all I remember is, friends flocking round
As I sat with his head 'twixt my knees on the ground;
And no voice but was praising this Roland of mine,
As I poured down his throat our last measure of wine,
Which (the burgesses voted by common consent)
Was no more than his due who brought good news from Ghent.

ROBERT BROWNING

After Blenheim

It was a summer evening,
  Old Kaspar's work was done,
And he before his cottage door
  Was sitting in the sun,
And by him sported on the green
His little grandchild Wilhelmine.

She saw her brother Peterkin
  Roll something large and round,
Which he beside the rivulet
  In playing there had found;
He came to ask what he had found,
That was so large, and smooth, and round.

Old Kaspar took it from the boy,
  Who stood expectant by;
And then the old man shook his head,
  And with a natural sigh,
''Tis some poor fellow's skull,' said he,
'Who fell in the great victory.

'I find them in the garden,
  For there's many here about;
And often when I go to plough,
  The ploughshare turns them out!
For many thousand men,' said he,
'Were slain in that great victory.'

'Now tell us what 'twas all about,'
  Young Peterkin he cries;
And little Wilhelmine looks up
  With wonder-waiting eyes;
'Now tell us all about the war,
And what they fought each other for.'

'It was the English,' Kaspar cried,
  'Who put the French to rout;
But what they fought each other for
  I could not well make out;

But everybody said,' quoth he,
'That 'twas a famous victory.

'My father lived at Blenheim then,
  Yon little stream hard by;
They burnt his dwelling to the ground,
  And he was forced to fly;
So with his wife and child he fled,
Nor had he where to rest his head.

'With fire and sword the country round
  Was wasted far and wide,
And many a childing mother then,
  And new-born baby died;
But things like that, you know, must be
At every famous victory.

'They say it was a shocking sight
  After the field was won;
For many thousand bodies here
  Lay rotting in the sun;
But things like that, you know, must be
After a famous victory.

'Great praise the Duke of Marlbro' won,
  And our good Prince Eugene.'
'Why, 'twas a very wicked thing!'
  Said little Wilhelmine.
'Nay...nay...my little girl,' quoth he,
'It was a famous victory.

'And everybody praised the Duke
  Who this great fight did win.'
'But what good came of it at last?'
  Quoth little Peterkin.
'Why, that I cannot tell,' said he,
'But 'twas a famous victory.'

ROBERT SOUTHEY

A Runnable Stag

When the pods went pop on the broom, green broom,
  And apples began to be golden-skinned,
We harboured a stag in the Priory coomb,
  And we feathered his trail up-wind, up-wind,
  We feathered his trail up-wind—
    A stag of warrant, a stag, a stag,
    A runnable stag, a kingly crop,
    Brow, bay and tray and three on top,
    A stag, a runnable stag.

Then the huntsman's horn rang yap, yap, yap,
  And 'Forwards' we heard the harbourer shout;
But 'twas only a brocket that broke a gap
  In the beechen underwood, driven out,
  From the underwood antlered out
    By warrant and might of the stag, the stag,
    The runnable stag, whose lordly mind
    Was bent on sleep, though beamed and tined
    He stood, a runnable stag.

So we tufted the covert till afternoon
  With Tinkerman's Pup and Bell-of-the-North;
And hunters were sulky and hounds out of tune
  Before we tufted the right stag forth,
  Before we tufted him forth,
    The stag of warrant, the wily stag,
    The runnable stag with his kingly crop,
    Brow, bay and tray and three on top,
    The royal and runnable stag.

It was Bell-of-the-North and Tinkerman's Pup
  That stuck to the scent till the copse was drawn.
'Tally ho! tally ho!' and the hunt was up,
  The tufters whipped and the pack laid on,
  The resolute pack laid on,

coomb] deep glen. Brow, bay and tray] the first, second and third antlers. brocket] stag in his second year. beamed and tined] the stag's antler consists of a 'beam' or stem and 'tines' or prongs.

And the stag of warrant away at last,
The runnable stag, the same, the same,
His hoofs on fire, his horns like flame,
A stag, a runnable stag.

'Let your gelding be: if you check or chide
He stumbles at once and you're out of the hunt;
For three hundred gentlemen, able to ride,
On hunters accustomed to bear the brunt,
Accustomed to bear the brunt,
Are after the runnable stag, the stag,
The runnable stag with his kingly crop,
Brow, bay and tray and three on top,
The right, the runnable stag.'

By perilous paths in coomb and dell,
The heather, the rocks, and the river-bed,
The pace grew hot, for the scent lay well,
And a runnable stag goes right ahead,
The quarry went right ahead—
Ahead, ahead, and fast and far;
His antlered crest, his cloven hoof,
Brow, bay and tray and three aloof,
The stag, the runnable stag.

For a matter of twenty miles and more,
By the densest hedge and the highest wall,
Through herds of bullocks he baffled the lore
Of harbourer, huntsman, hounds and all,
Of harbourer, hounds and all—
The stag of warrant, the wily stag,
For twenty miles, and five and five,
He ran, and he never was caught alive,
This stag, this runnable stag.

When he turned at bay in the leafy gloom,
In the emerald gloom where the brook ran deep,
He heard in the distance the rollers boom,
And he saw in a vision of peaceful sleep,
In a wonderful vision of sleep,

A stag of warrant, a stag, a stag,
A runnable stag in a jewelled bed,
Under the sheltering ocean dead,
A stag, a runnable stag.

So a fateful hope lit up his eye,
And he opened his nostrils wide again,
And he tossed his branching antlers high
As he headed the hunt down the Charlock glen,
As he raced down the echoing glen
For five miles more, the stag, the stag,
For twenty miles, and five and five,
Not to be caught now, dead or alive,
The stag, the runnable stag.

Three hundred gentlemen, able to ride,
Three hundred horses as gallant and free,
Beheld him escape on the evening tide,
Far out till he sank in the Severn Sea,
Till he sank in the depths of the sea—
The stag, the buoyant stag, the stag
That slept at last in a jewelled bed
Under the sheltering ocean spread,
The stag, the runnable stag.

JOHN DAVIDSON

Hart-leap Well

PART I

The Knight had ridden down from Wensley Moor
With the slow motion of a summer's cloud,
And now, as he approached a vassal's door,
'Bring forth another horse!' he cried aloud.

'Another horse!'—That shout the vassal heard
And saddled his best steed, a comely grey;
Sir Walter mounted him; he was the third
Which he had mounted on that glorious day.

Joy sparkled in the prancing courser's eyes;
The horse and horseman are a happy pair;
But, though Sir Walter like a falcon flies,
There is a doleful silence in the air.

A rout this morning left Sir Walter's hall,
That as they galloped made the echoes roar;
But horse and man are vanished, one and all;
Such race, I think, was never seen before.

Sir Walter, restless as a veering wind,
Calls to the few tired dogs that yet remain:
Blanch, Swift, and Music, noblest of their kind,
Follow, and up the weary mountain strain.

The Knight hallooed, he cheered and chid them on
With suppliant gestures and upbraidings stern;
But breath and eyesight fail; and, one by one,
The dogs are stretched among the mountain fern.

Where is the throng, the tumult of the race?
The bugles that so joyfully were blown?
—This chase it looks not like an earthly chase;
Sir Walter and the hart are left alone.

The poor hart toils along the mountain-side;
I will not stop to tell how far he fled,
Nor will I mention by what death he died;
But now the Knight beholds him lying dead.

Dismounting, then, he leaned against a thorn;
He had no follower, dog, nor man, nor boy:
He neither cracked his whip, nor blew his horn,
But gazed upon the spoil with silent joy.

Close to the thorn on which Sir Walter leaned
Stood his dumb partner in this glorious feat;
Weak as a lamb the hour that it is yeaned;
And white with foam as if with cleaving sleet.

Upon his side the hart was lying stretched:
His nostril touched a spring beneath a hill,
And with the last deep groan his breath had fetched
The waters of the spring were trembling still.

And now, too happy for repose or rest,
(Never had living man such joyful lot!)
Sir Walter walked all round, north, south, and west
And gazed and gazed upon that darling spot.

And climbing up the hill—(it was at least
Four roods of sheer ascent) Sir Walter found
Three several hoof-marks which the hunted beast
Had left imprinted on the grassy ground.

Sir Walter wiped his face, and cried, 'Till now
Such sight was never seen by human eyes:
Three leaps have borne him from this lofty brow
Down to the very fountain where he lies.

'I'll build a pleasure-house upon this spot,
And a small arbour, made for rural joy;
'Twill be the traveller's shed, the pilgrim's cot,
A place of love for damsels that are coy.

'A cunning artist will I have to frame
A basin for that fountain in the dell!
And they who do make mention of the same,
From this day forth, shall call it HART-LEAP WELL.

'And, gallant stag! to make thy praises known,
Another monument shall here be raised;
Three several pillars, each a rough-hewn stone,
And planted where thy hoofs the turf have grazed.

'And in the summer-time, when days are long,
I will come hither with my paramour;
And with the dancers and the minstrel's song
We will make merry in that pleasant bower.

'Till the foundations of the mountains fail
My mansion with its arbour shall endure;—
The joy of them who till the fields of Swale,
And them who dwell among the woods of Ure!'

Then home he went, and left the hart stone-dead,
With breathless nostrils stretched above the spring.
—Soon did the Knight perform what he had said;
And far and wide the fame thereof did ring.

Ere thrice the moon into her port had steered,
A cup of stone received the living well;
Three pillars of rude stone Sir Walter reared,
And built a house of pleasure in the dell.

And, near the fountain, flowers of stature tall
With trailing plants and trees were intertwined,—
Which soon composed a little sylvan hall,
A leafy shelter from the sun and wind.

And thither, when the summer days were long,
Sir Walter led his wondering paramour;
And with the dancers and the minstrel's song
Made merriment within that pleasant bower.

The Knight, Sir Walter, died in course of time,
And his bones lie in his paternal vale.—
But there is matter for a second rhyme,
And I to this would add another tale.

PART II

The moving accident is not my trade;
To freeze the blood I have no ready arts:
'Tis my delight, alone in summer shade,
To pipe a simple song for thinking hearts.

As I from Hawes to Richmond did repair,
It chanced that I saw standing in a dell
Three aspens at three corners of a square;
And one, not four yards distant, near a well.

What this imported I could ill divine:
And, pulling now the rein my horse to stop,
I saw three pillars standing in a line,—
The last stone-pillar on a dark hill-top.

The trees were grey, with neither arms nor head;
Half wasted the square mound of tawny green;
So that you just might say, as then I said,
'Here in old time the hand of man hath been.'

I looked upon the hill both far and near,
More doleful place did never eye survey;
It seemed as if the spring-time came not here,
And Nature here were willing to decay.

I stood in various thoughts and fancies lost,
When one, who was in shepherd's garb attired,
Came up the hollow:—him did I accost,
And what this place might be I then inquired.

The Shepherd stopped, and that same story told
Which in my former rhyme I have rehearsed.
'A jolly place,' said he, 'in times of old!
But something ails it now: the spot is curst.

'You see these lifeless stumps of aspen wood—
Some say that they are beeches, others elms—
These were the bower; and here a mansion stood,
The finest palace of a hundred realms!

'The arbour does its own condition tell;
You see the stones, the fountain, and the stream;
But as to the great Lodge! you might as well
Hunt half a day for a forgotten dream.

'There's neither dog nor heifer, horse nor sheep,
Will wet his lips within that cup of stone;
And oftentimes, when all are fast asleep,
This water doth send forth a dolorous groan.

'Some say that here a murder has been done,
And blood cries out for blood: but, for my part,
I've guessed, when I've been sitting in the sun,
That it was all for that unhappy hart.

'What thoughts must through the creature's brain have past!
Even from the topmost stone, upon the steep,
Are but three bounds—and look, Sir, at this last—
O Master! it has been a cruel leap.

'For thirteen hours he ran a desperate race;
And in my simple mind we cannot tell
What cause the hart might have to love this place,
And come and make his death-bed near the well.

'Here on the grass perhaps asleep he sank,
Lulled by the fountain in the summer-tide;
This water was perhaps the first he drank
When he had wandered from his mother's side.

'In April here, beneath the flowering thorn,
He heard the birds their morning carols sing;
And he perhaps, for aught we know, was born
Not half a furlong from that self-same spring.

'Now, here is neither grass nor pleasant shade;
The sun on drearier hollow never shone;
So will it be, as I have often said,
Till trees, and stones, and fountain, all are gone.'

'Grey-headed shepherd, thou hast spoken well;
Small difference lies between thy creed and mine:
This beast not unobserved by Nature fell;
His death was mourned by sympathy divine.

'The Being that is in the clouds and air,
That is in the green leaves among the groves,
Maintains a deep and reverential care
For the unoffending creatures whom he loves.

'The pleasure-house is dust:—behind, before,
This is no common waste, no common gloom;
But Nature, in due course of time, once more
Shall here put on her beauty and her bloom.

'She leaves these objects to a slow decay,
That what we are, and have been, may be known;
But at the coming of the milder day
These monuments shall all be overgrown.

'One lesson, shepherd, let us two divide,
Taught both by what she shows, and what conceals;
Never to blend our pleasure or our pride
With sorrow of the meanest thing that feels.'

WILLIAM WORDSWORTH

The Bull

See an old unhappy bull,
Sick in soul and body both,
Slouching in the undergrowth
Of the forest beautiful,
Banished from the herd he led,
Bulls and cows a thousand head.

Cranes and gaudy parrots go
Up and down the burning sky;
Tree-top cats purr drowsily
In the dim-day green below;
And troops of monkeys, nutting, some,
All disputing, go and come;

And things abominable sit
Picking offal buck or swine,
On the mess and over it
Burnished flies and beetles shine,
And spiders big as bladders lie
Under hemlocks ten foot high;

And a dotted serpent curled
Round and round and round a tree,
Yellowing its greenery,
Keeps a watch on all the world,
All the world and this old bull
In the forest beautiful.

Bravely by his fall he came:
One he led, a bull of blood
Newly come to lustihood,
Fought and put his prince to shame,
Snuffed and pawed the prostrate head
Tameless even while it bled.

There they left him, every one,
Left him there without a lick,
Left him for the birds to pick,
Left him there for carrion,
Vilely from their bosom cast
Wisdom, worth, and love at last.

When the lion left his lair
And roared his beauty through the hills,
And the vultures pecked their quills
And flew into the middle air,
Then this prince no more to reign
Came to life and lived again.

He snuffed the herd in far retreat,
He saw the blood upon the ground,
And snuffed the burning airs around
Still with beevish odours sweet,
While the blood ran down his head
And his mouth ran slaver red.

Pity him, this fallen chief,
All his splendour, all his strength,
All his body's breadth and length
Dwindled down with shame and grief,
Half the bull he was before,
Bones and leather, nothing more.

See him standing dewlap-deep
In the rushes at the lake,
Surly, stupid, half asleep,
Waiting for his heart to break
And the birds to join the flies
Feasting at his bloodshot eyes,—

Standing with his head hung down
In a stupor, dreaming things:
Green savannas, jungles brown,
Battlefields and bellowings,
Bulls undone and lions dead
And vultures flapping overhead.

Dreaming things: of days he spent
With his mother gaunt and lean
In the valley warm and green,
Full of baby wonderment,
Blinking out of silly eyes
At a hundred mysteries;

Dreaming over once again
How he wandered with a throng
Of bulls and cows a thousand strong,
Wandered on from plain to plain,
Up the hill and down the dale,
Always at his mother's tail.

How he lagged behind the herd,
Lagged and tottered, weak of limb,
And she turned and ran to him
Blaring at the loathly bird
Stationed always in the skies,
Waiting for the flesh that dies.

Dreaming maybe of a day
When her drained and drying paps
Turned him to the sweets and saps,
Richer fountains by the way,
And she left the bull she bore
And he looked to her no more;

And his little frame grew stout,
And his little legs grew strong,
And the way was not so long;
And his little horns came out,
And he played at butting trees,
And boulder-stones and tortoises,

Joined a game of knobby skulls
With the youngsters of his year,
And the other little bulls,
Learning both to bruise and bear,
Learning how to stand a shock
Like a little bull of rock.

Dreaming of a day less dim,
Dreaming of a time less far,
When the faint but certain star
Of destiny burned clear for him,
And a fierce and wild unrest
Broke the quiet of his breast,

And the gristles of his youth
Hardened in his comely pow,
And he came to fighting growth,
Beat his bull and won his cow,
And flew his tail and trampled off
Past the tallest, vain enough,

And curved about in splendour full
And curved again and snuffed the airs
As who should say, Come out who dares!
And all beheld a bull, a Bull,
And knew that here was surely one
That backed for no bull, fearing none.

And the leader of the herd
Looked and saw, and beat the ground,
And shook the forest with his sound,
Bellowed at the loathly bird
Stationed always in the skies,
Waiting for the flesh that dies.

Dreaming, this old bull forlorn,
Surely dreaming of the hour
When he came to sultan power,
And they owned him master-horn,
Chiefest bull of all among
Bulls and cows a thousand strong.

And in all the tramping herd
Not a bull that barred his way,
Not a cow that said him nay,
Not a bull or cow that erred
In the furnace of his look
Dared a second, worse rebuke;

Not in all the forest wide,
Jungle, thicket, pasture, fen,
Not another dared him then,
Dared him and again defied;
Not a sovereign buck or boar
Came a second time for more.

Not a serpent that survived
Once the terrors of his hoof
Risked a second time reproof,
Came a second time and lived,
Not a serpent in its skin
Came again for discipline;

Not a leopard bright as flame,
Flashing fingerhooks of steel,
That a wooden tree might feel,
Met his fury once and came
For a second reprimand,
Not a leopard in the land;

Not a lion of them all,
Not a lion of the hills,
Hero of a thousand kills,
Dared a second fight and fall,
Dared that ram terrific twice,
Paid a second time the price.

Pity him, this dupe of dream,
Leader of the herd again
Only in his daft old brain,
Once again the bull supreme
And bull enough to bear the part
Only in his tameless heart.

Pity him that he must wake;
Even now the swarm of flies
Blackening his bloodshot eyes
Bursts and blusters round the lake,
Scattered from the feast half-fed,
By great shadows overhead;

And the dreamer turns away
From his visionary herds
And his splendid yesterday,
Turns to meet the loathly birds
Flocking round him from the skies,
Waiting for the flesh that dies.

RALPH HODGSON

Fidelity

A barking sound the shepherd hears,
A cry as of a dog or fox;
He halts—and searches with his eyes
Among the scattered rocks:
And now at distance can discern
A stirring in a brake of fern;
And instantly a dog is seen,
Glancing through that covert green.

The dog is not of mountain breed;
Its motions, too, are wild and shy;
With something, as the shepherd thinks,
Unusual in its cry:

Nor is there any one in sight
All round, in hollow or on height;
Nor shout, nor whistle strikes his ear;
What is the creature doing here?

It was a cove, a huge recess,
That keeps, till June, December's snow;
A lofty precipice in front,
A silent tarn below!
Far in the bosom of Helvellyn,
Remote from public road or dwelling,
Pathway, or cultivated land;
From trace of human foot or hand.

There sometimes doth a leaping fish
Send through the tarn a lonely cheer;
The crags repeat the raven's croak,
In symphony austere;
Thither the rainbow comes—the cloud—
And mists that spread the flying shroud;
And sunbeams; and the sounding blast,
That, if it could, would hurry past;
But that enormous barrier holds it fast.

Not free from boding thoughts, a while
The shepherd stood; then makes his way
O'er rocks and stones, following the dog
As quickly as he may;
Nor far had gone before he found
A human skeleton on the ground;
The appalled discoverer with a sigh
Looks round, to learn the history.

From those abrupt and perilous rocks
The man had fallen, that place of fear!
At length upon the shepherd's mind
It breaks, and all is clear:
He instantly recalled the name,
And who he was, and whence he came;

Remembered, too, the very day
On which the traveller passed this way.

But hear a wonder, for whose sake
This lamentable tale I tell!
A lasting monument of words
This wonder merits well.
The dog, which still was hovering nigh,
Repeating the same timid cry,
This dog had been through three months' space
A dweller in that savage place.

Yes, proof was plain that, since the day
When this ill-fated traveller died,
The dog had watched about the spot,
Or by his master's side:
How nourished here through such long time
He knows, who gave that love sublime;
And gave that strength of feeling, great
Above all human estimate!

WILLIAM WORDSWORTH

An Elegy on the Death of a Mad Dog

Good people all, of every sort,
  Give ear unto my song;
And if you find it wondrous short,
  It cannot hold you long.

In Islington there was a man,
  Of whom the world might say,
That still a godly race he ran,
  Whene'er he went to pray.

A kind and gentle heart he had,
  To comfort friends and foes;
The naked every day he clad,
  When he put on his clothes.

And in that town a dog was found,
As many dogs there be,
Both mongrel, puppy, whelp, and hound,
And curs of low degree.

This dog and man at first were friends;
But when a pique began,
The dog, to gain his private ends,
Went mad, and bit the man.

Around from all the neighbouring streets
The wondering neighbours ran,
And swore the dog had lost his wits,
To bite so good a man.

The wound it seem'd both sore and sad
To every Christian eye;
And while they swore the dog was mad,
They swore the man would die.

But soon a wonder came to light,
That show'd the rogues they lied;
The man recover'd of the bite,
The dog it was that died.

OLIVER GOLDSMITH

The Retired Cat

A poet's cat, sedate and grave,
As poet well could wish to have,
Was much addicted to inquire
For nooks, to which she might retire,
And where, secure as mouse in chink,
She might repose, or sit and think.
I know not where she caught the trick—
Nature perhaps herself had cast her
In such a mould PHILOSOPHIQUE,
Or else she learn'd it of her master.
Sometimes ascending, debonair,
An apple-tree or lofty pear,

Lodg'd with convenience in the fork,
She watched the gard'ner at his work;
Sometimes her ease and solace sought
In an old empty wat'ring pot,
There wanting nothing, save a fan,
To seem some nymph in her sedan,
Apparell'd in exactest sort,
And ready to be borne to court.
But love of change it seems has place
Not only in our wiser race;
Cats also feel as well as we
That passion's force, and so did she.
Her climbing, she began to find,
Expos'd her too much to the wind,
And the old utensil of tin
Was cold and comfortless within:
She therefore wish'd, instead of those,
Some place of more serene repose,
Where neither cold might come, nor air
Too rudely wanton with her hair,
And sought it in the likeliest mode
Within her master's snug abode.
A draw'r,—it chanc'd, at bottom lin'd
With linen of the softest kind,
With such as merchants introduce
From India, for the ladies' use,—
A draw'r impending o'er the rest,
Half open in the topmost chest,
Of depth enough, and none to spare,
Invited her to slumber there.
Puss with delight beyond expression,
Survey'd the scene, and took possession.
Recumbent at her ease ere long,
And lull'd by her own humdrum song,
She left the cares of life behind,
And slept as she would sleep her last,
When in came, housewifely inclin'd,
The chambermaid, and shut it fast,

By no malignity impell'd,
But all unconscious whom it held.
Awaken'd by the shock (cried puss)
Was ever cat attended thus!
The open draw'r was left, I see,
Merely to prove a nest for me,
For soon as I was well compos'd,
Then came the maid, and it was closed:
How smooth these 'kerchiefs, and how sweet,
O what a delicate retreat!
I will resign myself to rest
Till Sol, declining in the west,
Shall call to supper; when, no doubt,
Susan will come and let me out.
The evening came, the sun descended,
And puss remain'd still unattended.
The night roll'd tardily away,
(With her indeed 'twas never day)
The sprightly morn her course renew'd,
The evening gray again ensued,
And puss came into mind no more
Than if entomb'd the day before.
With hunger pinch'd, and pinch'd for room,
She now presag'd approaching doom,
Nor slept a single wink, or purr'd,
Conscious of jeopardy incurr'd.
That night, by chance, the poet watching,
Heard an inexplicable scratching;
His noble heart went pit-a-pat,
And to himself he said—what's that?
He drew the curtain at his side,
And forth he peep'd, but nothing spied.
Yet, by his ear directed, guess'd
Something imprison'd in the chest,
And doubtful what, with prudent care,
Resolv'd it should continue there.
At length a voice, which well he knew,
A long and melancholy mew,

Saluting his poetic ears,
Consol'd him, and dispell'd his fears;
He left his bed, he trod the floor,
He 'gan in haste the draw'rs explore,
The lowest first, and without stop,
The rest in order to the top.
For 'tis a truth well known to most,
That whatsoever thing is lost,
We seek it, ere it come to light,
In ev'ry cranny but the right.
Forth skipp'd the cat; not now replete
As erst with airy self-conceit,
Nor in her own fond apprehension
A theme for all the world's attention,
But modest, sober, cur'd of all
Her notions hyperbolical,
And wishing for a place of rest
Any thing rather than a chest:
Then stept the poet into bed,
With this reflexion in his head:

MORAL

Beware of too sublime a sense
Of your own worth and consequence!
The man who dreams himself so great,
And his importance of such weight,
That all around, in all that's done,
Must move and act for him alone,
Will learn, in school of tribulation,
The folly of his expectation.

WILLIAM COWPER

The Colubriad

Close by the threshold of a door nail'd fast
Three kittens sat: each kitten look'd aghast.
I, passing swift and inattentive by,
At the three kittens cast a careless eye;

Not much concerned to know what they did there,
Not deeming kittens worth a poet's care.
But presently a loud and furious hiss
Caused me to stop, and to exclaim—what's this?
When, lo! upon the threshold met my view,
With head erect, and eyes of fiery hue,
A viper, long as Count de Grasse's queue.
Forth from his head his forked tongue he throws,
Darting it full against a kitten's nose;
Who having never seen in field or house
The like, sat still and silent, as a mouse:
Only, projecting with attention due
Her whisker'd face, she ask'd him—who are you?
On to the hall went I, with pace not slow,
But swift as lightning, for a long Dutch hoe;
With which well arm'd I hasten'd to the spot
To find the viper. But I found him not,
And, turning up the leaves and shrubs around,
Found only, that he was not to be found.
But still the kittens, sitting as before,
Sat watching close the bottom of the door.
I hope—said I—the villain I would kill
Has slipt between the door and the door's sill;
And if I make despatch, and follow hard,
No doubt but I shall find him in the yard:—
For long ere now it should have been rehears'd
'Twas in the garden that I found him first.
E'en there I found him; there the full-grown cat
His head with velvet paw did gently pat,
As curious as the kittens erst had been
To learn what this phenomenon might mean.
Fill'd with heroic ardour at the sight,
And fearing every moment he would bite,
And rob our household of our only cat
That was of age to combat with a rat,
With out-stretch'd hoe I slew him at the door,
And taught him NEVER TO COME THERE NO MORE.

WILLIAM COWPER

SECTION 4

THE POET AND THE MODERN WORLD

Primitive people, as we have seen, believed that they must please the forces of nature, which were to them such deities as the god of thunder and the god of the wind, in order to grow their crops safely and so keep themselves alive in a hostile world. They lived in constant fear of drought and storms and floods. They rejoiced in the coming of Spring and in the warm sunshine. Accustomed as they were to thinking of the world of nature as though it were peopled with spirits having human virtues and human vices, they strove in their poetry and art and in their way of life to establish a harmony of human culture with their natural environment. This had a great influence upon mankind.

All this has now changed. The modern world is a world of machines, and we are trying to adjust ourselves to a complete revolution in living.

At first the new inventions were hated and feared just as the forces of nature were feared. A poet would write about those 'dark Satanic mills', justified perhaps by the terrible conditions in factories in the early nineteenth century.

Slowly this attitude was modified. Men grew more sympathetic in their reactions, and poets tried to accept this new force in a changing world, declaring the machines to be

> greater than the People or the Kings.

Poets now wrote in praise, speaking as of a living thing:

> What nudity as beautiful as this
> Obedient monster purring at its toil

and pointing to those

naked iron muscles dripping oil.

Here the machine is the servant of man, willing and tireless, with whose help would come an age of peace and plenty. When instead came terrible, devastating wars, poverty, unemployment, slums, a new ugliness and a ruined countryside, poets were the first to warn mankind that these inventions, the products of man's amazing skill and ingenuity might well, through our own folly, enslave and then destroy us.

When we are carried away by enthusiasm for the wonderful achievements of our modern world, we should be prepared to pause and listen to such poets, because out of their sincere and intense feelings they may be speaking to us as true prophets.

Portrait of a Machine

What nudity as beautiful as this
Obedient monster purring at its toil;
Those naked iron muscles dripping oil,
And the sure-fingered rods that never miss?
This long and shining flank of metal is
Magic that greasy labour cannot spoil;
While this vast engine that could rend the soil
Conceals its fury with a gentle hiss.

It does not vent its loathing, it does not turn
Upon its makers with destroying hate.
It bears a deeper malice; lives to earn
Its master's bread and laughs to see this great
Lord of the earth, who rules but cannot learn,
Become the slave of what his slaves create.

LOUIS UNTERMEYER

The Secret of the Machines

We were taken from the ore-bed and the mine,
  We were melted in the furnace and the pit—
We were cast and wrought and hammered to design,
  We were cut and filed and tooled and gauged to fit.
Some water, coal, and oil is all we ask,
  And a thousandth of an inch to give us play,
And now if you will set us to our task,
  We will serve you four and twenty hours a day!

    We can pull and haul and push and lift and drive,
    We can print and plough and weave and heat and light,
    We can run and jump and swim and fly and dive,
    We can see and hear and count and read and write!

Would you call a friend from half across the world?
  If you'll let us have his name and town and state,
You shall see and hear your crackling question hurled
  Across the arch of heaven while you wait.
Has he answered? Does he need you at his side?
  You can start this very evening if you choose,
And take the Western Ocean in the stride
  Of thirty thousand horses and some screws!

    The boat-express is waiting your command!
    You will find the *Mauretania* at the quay,
    Till her captain turns the lever 'neath his hand,
    And the monstrous nine-decked city goes to sea.

Do you wish to make the mountains bare their head
  And lay their new-cut forests at your feet?
Do you want to turn a river in its bed,
  And plant a barren wilderness with wheat?
Shall we pipe aloft and bring you water down
  From the never-failing cisterns of the snows,
To work the mills and tramways in your town,
  And irrigate your orchards as it flows?

It is easy! Give us dynamite and drills!
Watch the iron-shouldered rocks lie down and quake
As the thirsty desert-level floods and fills,
And the valley we have dammed becomes a lake!

But remember, please, the Law by which we live,
We are not built to comprehend a lie,
We can neither love nor pity nor forgive,
If you make a slip in handling us you die!
We are greater than the Peoples or the Kings—
Be humble, as you crawl beneath our rods!—
Our touch can alter all created things,
We are everything on earth—except The Gods!

Though our smoke may hide the Heavens from your eyes,
It will vanish and the stars will shine again,
Because, for all our power and weight and size,
We are nothing more than children of your brain!

RUDYARD KIPLING

The Pigeon

Throb, throb from the mixer
Spewing out concrete.
And at the heads of the cables
Stand the serpent-warders,
Sweating and straining,
Thrusting those cruel mouths to their prey.

Hark how the steel tongues hiss
As they stab.
The men sway under the effort,
And their eyes are bloodshot with the din,
The clatter that shatters the brain.
Throb, throb from the mixer
Spewing out concrete.

The crowd stands by
Watching the smoothers;
Fascinated by the flat, wet levels
Of newlaid cement.

See how those curdled lakes
Glisten under the sky,
Virginal.

Then the dusty air suddenly divides,
And a pigeon from a plane-tree
Flutters down to bathe its wings in that mirage of water.

But deceived, and angry,
Bewildered by the din,
The throb, throb from the mixer
Spewing out concrete,
It backs upon its wing,
Threshes air, and is gone.

But there, in the deflowered bed,
Is the seal of its coral foot,
Set till rocks crumble.

RICHARD CHURCH

To a Telegraph Pole

You should be done with blossoming by now.
Yet here are leaves closer than any bough
That welcomes ivy. True, you were a tree
And stood with others in a marching line,
Less regular than this, of spruce and pine,
And boasted branches rather than a trunk.
This is your final winter, all arms shrunk
To one cross-bar bearing haphazardly
Four rusty strands. You cannot hope to feel
The electric sap run through those veins of steel.
The birds know this; the birds have hoodwinked you,
Crowding about you as they used to do.
The rainy robins huddled on your wire
And those black birds with shoulders dipped in fire
Have made you dream these vines; these tendrils are
A last despair in green, familiar
To derelicts of earth as well as sea.
Do not believe them, there is mockery

In their cool little jets of song. They know
What everyone but you learned long ago:
The stream of stories humming through your head
Is not your own. You dream. But you are dead.

LOUIS UNTERMEYER

The Line-Gang

Here come the line-gang pioneering by.
They throw a forest down less cut than broken.
They plant dead trees for living, and the dead
They string together with a living thread.
They string an instrument against the sky
Wherein words, whether beaten out or spoken,
Will run as hushed as when they were a thought.
But in no hush they string it: they go past
With shouts afar to pull the cable taut,
To hold it hard until they make it fast,
To ease away—they have it. With a laugh
And oath of towns that set the wild at naught,
They bring the telephone and the telegraph.

ROBERT FROST

Factory Windows are always Broken

Factory windows are always broken.
Somebody's always throwing bricks,
Somebody's always heaving cinders,
Playing ugly Yahoo tricks.

Factory windows are always broken.
Other windows are let alone.
No one throws through the chapel-window
The bitter, snarling derisive stone.

Factory windows are always broken.
Something or other is going wrong.
Something is rotten—I think, in Denmark.
End of the factory-window song.

VACHEL LINDSAY

Prelude

The winter evening settles down
With smells of steaks in passageways.
Six o'clock.
The burnt-out ends of smoky days.
And now a gusty shower wraps
The grimy scraps
Of withered leaves about your feet
And newspapers from vacant lots;
The showers beat
On broken blinds and chimney-pots,
And at the corner of the street
A lonely cab-horse steams and stamps.
And then the lighting of the lamps.

T. S. ELIOT

He Will Watch the Hawk

He will watch the hawk with an indifferent eye
Or pitifully;
Nor on those eagles that so feared him, now
Will strain his brow;
Weapons men use, stone, sling and strong-thewed bow
He will not know.

This aristocrat, superb of all instinct,
With death close linked
Had paced the enormous cloud, almost had won
War on the sun;
Till now, like Icarus mid-ocean-drowned,
Hands, wings, are found.

STEPHEN SPENDER

Naming of Parts

To-day we have naming of parts. Yesterday,
We had daily cleaning. And to-morrow morning,
We shall have what to do after firing. But to-day,
To-day we have naming of parts. Japonica
Glistens like coral in all of the neighbouring gardens,
    And to-day we have naming of parts.

This is the lower sling swivel. And this
Is the upper sling swivel, whose use you will see,
When you are given your slings. And this is the piling swivel,
Which in your case you have not got. The branches
Hold in the gardens their silent, eloquent gestures,
    Which in our case we have not got.

This is the safety-catch, which is always released
With an easy flick of the thumb. And please do not let me
See anyone using his finger. You can do it quite easy
If you have any strength in your thumb. The blossoms
Are fragile and motionless, never letting anyone see
    Any of them using their finger.

And this you can see is the bolt. The purpose of this
Is to open the breech, as you see. We can slide it
Rapidly backwards and forwards: we call this
Easing the spring. And rapidly backwards and forwards
The early bees are assaulting and fumbling the flowers:
    They call it easing the Spring.

They call it easing the Spring: it is perfectly easy
If you have any strength in your thumb: like the bolt,
And the breech, and the cocking-piece, and the point of balance,
Which in our case we have not got; and the almond-blossom
Silent in all of the gardens and the bees going backwards and
        forwards,
    For to-day we have naming of parts.

HENRY REED

Parliament Hill Fields

Rumbling under blackened girders, Midland bound for Cricklewood,
Puffed its sulphur to the sunset where that Land of Laundries stood.
Rumble under, thunder over, train and tram alternate go,
Shake the floor and smudge the ledger, Charrington, Sells, Dale
 and Co.,
Nuts and nuggets in the window, trucks along the lines below.

When the Bon Marché was shuttered, when the feet were hot and
 tired,
Outside Charrington's we waited, by the 'STOP HERE IF REQUIRED',
Launched aboard the shopping basket, sat precipitately down,
Rocked past Zwanziger the baker's, and the terrace blackish brown
And the curious Anglo-Norman parish church of Kentish Town.

Till the tram went over thirty, sighting terminus again,
Past municipal lawn tennis and the bobble-hanging plane;
Soft the light suburban evening caught our ashlar-speckled spire,
Eighteen sixty Early English, as the mighty elms retire
Either side of Brookfield Mansions flashing fine French-window fire.

Oh the after-tram-ride quiet, when we heard a mile beyond,
Silver music from the bandstand, barking dogs by Highgate Pond;
Up the hill where stucco houses in Virginia creeper drown—
And my childish wave of pity, seeing children carrying down
Sheaves of drooping dandelions to the courts of Kentish Town.

JOHN BETJEMAN

To Iron-Founders and Others

When you destroy a blade of grass
You poison England at her roots:
Remember no man's foot can pass
Where evermore no green life shoots.

You force the birds to wing too high
Where your unnatural vapours creep:
Surely the living rocks shall die
When birds no rightful distance keep.

You have brought down the firmament
And yet no heaven is more near;
You shape huge deeds without event,
And half-made men believe and fear.

Your worship is your furnaces,
Which, like old idols, lost obscenes,
Have molten bowels; your vision is
Machines for making more machines.

O, you are busied in the night,
Preparing destinies of rust;
Iron misused must turn to blight
And dwindle to a tetter'd crust.

The grass, forerunner of life, has gone,
But plants that spring in ruins and shards
Attend until your dream is done:
I have seen hemlock in your yards.

The generations of the worm
Know not your loads piled on their soil;
Their knotted ganglions shall wax firm
Till your strong flagstones heave and toil.

When the old hollow'd earth is crack'd,
And when, to grasp more power and feasts,
Its ores are emptied, wasted, lack'd,
The middens of your burning beasts

Shall be raked over till they yield
Last priceless slags for fashionings high,
Ploughs to wake grass in every field,
Chisels men's hands to magnify.

GORDON BOTTOMLEY

To Some Builders of Cities

You have thrust Nature out, to make
A wilderness where nothing grows
But forests of unbudding stone

(The sparrow's lonely for his boughs);
You fling up citadels to stay
The soft invasion of the rose.

But though you put the Earth in thrall
And ransack all her fragrant dowers,
Her old accomplice, Heaven, will plot
To take with stars your roofs and towers;
And neither stone nor steel can foil
That silver strategy of flowers.

STANLEY SNAITH

Beleaguered Cities

Build your houses, build your houses, build your towns,
  Fell the woodland, to a gutter turn the brook,
Pave the meadows, pave the meadows, pave the downs,
  Plant your bricks and mortar where the grasses shook,
  The wind-swept grasses shook.
Build, build your Babels black against the sky—
But mark yon small green blade, your stones between,
  The single spy
Of that uncounted host you have outcast;
For with their tiny pennons waving green
  They shall storm your streets at last.

Build your houses, build your houses, build your slums,
  Drive your drains where once the rabbits used to lurk,
Let there be no song there save the wind that hums
  Through the idle wires while dumb men tramp to work,
  Tramp to their idle work.
Silent the siege; none notes it; yet one day
Men from your walls shall watch the woods once more
  Close round their prey.
Build, build the ramparts of your giant-town;
Yet they shall crumble to the dust before
  The battering thistle-down.

F. L. LUCAS

Time, Gentlemen, Time!

O would not Life be charming
Could we get rid of clocks,
The still ones and alarming
That break on sleep with shocks,

Then it would be respected
And worthier far of Man
Than when by springs directed
From gold or a tin can.

Why should Man's life be reckoned
By anything so queer
As that which splits the second
But cannot tell the year?

If we got rid of watches
The trains would cease to run,
We could not fight a battle ship
Or aim a battle gun,

Nor tune the little engines
Which fill the towns with fumes
And send men with a vengeance
(Quite rightly) to their tombs.

If we got rid of watches
And wanted to approach
The pallid peopled cities
We'd have to hire a coach

And guard, who, to arouse us,
So hardy in the morn,
Outside the licensed houses
Would blow a long bright horn.

Our stars know naught of watches,
There's not a wind that wists
Of mischief that Time hatches
When handcuffed to our wrists.

No wonder stars are winking,
  No wonder heaven mocks
At men who cease from drinking
  Good booze because of clocks!

'Twould make a devil chortle
  To see how all the clean
Free souls God made immortal
  Must march to a machine.

It makes me wonder whether
  In this grim pantomime
Did fiend or man first blether:
  'Time, Gentlemen, Time!'

We must throw out the timing
  That turns men into gnomes,
Of piece-work and of miming
  That fills the mental homes.

We must get rid of errors,
  And tallies and time checks,
And all the slavish terrors
  That turn men into wrecks.

They have not squared the circle,
  They have not cubed the sphere,
Their calendars all work ill
  Corrected by 'leap' year.

But we should all be leaping
  As high as hollyhocks
Did we desist from keeping
  Our trysts with slaves of clocks.

How should we tell the seconds?
  The time a blackbird takes,
To screech across a laneway,
  And dive into the brakes.

How should we tell the minutes?
The time it takes to swipe
A lonely pint of Guinness,
Or load a friendly pipe.

O make the heart Time's measure
Because, the more it beats,
The more Life fills with pleasure
With songs or sturdy feats;

Our clocks our lives are cheating
They go, and ground we give;
The higher the heart's beating
The higher then we live.

O. ST J. GOGARTY

Clock

When first you learn to read a clock
That moment you are in a snare,
Doomed for the rest of life to stand
A victim to that patient hand.

The large round eyes of time begin to stare;
The voice of time,
With tick and tock,
Beats like a heart against your ear.

Now all the clocks form close about,
And from the middle of that ring
You crave to find one passage out
In horror what their time may bring.
And is there no escape outside the circle
Where everything you do is overlooked?

I'd like to stare them through the eyes,
And see beyond that moony dial:
For backward from the axis of a clock,
Like gossamer at first,
Tight-braced strands, and cords becoming chains,

Lead, climb, and spread themselves away in space;
So It, their intimate converging place,
Acquires gigantic intricate communions,
Copious relation to forces beyond forces,
(Cool and placid though it look).
Away and away beyond it, range on range,
In all their tortures elemental courses,
The hidden worlds pursue their time and change;
Are, and then are no more,
Then are again—while we,
Crouched near their ticking dials, faintly guess,
And, as when listening to a far-off ocean,
Hear more, hear less,
Then often not at all,
And visualize the foamy green commotion
Of the great roaring waves that break and fall.

HAROLD MONRO

PART IV

SECTION I

THE MUSIC OF POETRY

The dancing of the tribe to the beat of drums, the repeated shout of victory or defeat, the rhythm of stamping feet were the first cause of poetry. A regular rhythm from the feet of savages became in time the feet of poetic metre. The chorus of voices keeping time to native music led gradually to more sustained and complicated choruses. These took on a definite poetic rhythm, a pattern of sound. Metre is that regular mechanical pattern built upon the stressed and unstressed syllables; the accent or stress of the voice corresponding to the beats in music or the beat of one's feet in dancing:

> The dóuble dóuble dóuble beát
> Of the thúndering drúm.

There is an infinite variety of such patterns and the poet uses a particular pattern to suit his mood. The patterns depend upon the arrangement of the stressed and unstressed syllables, the length of the line, the number of lines in a verse; on the arrangement of the rhyming words, and the rhythm given to the poem as a whole. Alliteration, the recurrence of the same letter or sound in close succession, is frequently used to create a particular sound effect:

> And on a sudden, lo! the level lake,
> And the long glories of the winter moon.

Rhyme is the repetition of syllables having the same sound, as in 'chair' and 'fair', 'write' and 'smite'; 'ending' and 'bending' is a double or feminine rhyme. Assonance is the correspondence of vowel sounds in two syllables without the identity of consonant sounds which would make a rhyme: 'drown', 'crowd'; 'side' and 'write'. Rhyming words usually come at the end of the line. Occasionally two words rhyme within the same line. This is known as medial rhyme:

Pack, clouds, *away*! and welcome, *day*!

Rhymes help to create the musical pattern by building up a tide of expectation. The ear begins to expect the recurrence of a particular sound and is pleased to recognize it when it comes. Onomatopoeia, the tendency in words to echo the meaning by the actual sound not only supports the sense but greatly enhances the musical effect. With rhyme, medial rhyme, assonance, alliteration, onomatopoeia there are sound echoes playing all over a poem. This frequent repetition of sound is common in poetry:

O hárk, O héar! how thín and cléar,
And thínner, cléarer, fárther góing!
O swéet and fár, from clíff and scár
The hórns of Élfland fáintly blówing!
Blów, let us héar the púrple gléns replýing:
Blów, búgle; ánswer, échoes, dýing, dýing, dýing.

Here the regular mechanical pattern which underlies rhythm is a succession of iambuses, an iambus being one unstressed syllable followed by a stressed syllable (dĕspáir). The poet has varied the metrical pattern in the last two lines, inverting the stresses with the last three words to give a succession of trochees, dýiňg, and the effect of an echo fading in the distance. The more unstressed syllables there are, the swifter is the movement of the line.

The Ăssýr | ĭăn căme dówn | lĭke thĕ wólf | ŏn thĕ fóld |

is anapaestic, the line consisting of four anapaests, two unstressed and one stressed syllable (sĕrĕnáde).

/ ⏑⏑ / ⏑⏑ / ⏑ ⏑ /
Merrily | merrily | shall I live | now

is dactyllic, the line consisting of three dactyls, one stressed and two unstressed syllables (trémŭloŭs).

This ebb and flow of sound, often with the repetition of word or phrase, makes the musical pattern. Now just as the slow beat of the native drums could put the audience into a trance-like state, while the rapid beating could stir to a frenzy, so the rhythmic sound of poetry can both excite your imagination, rouse your emotions, and also lull you, and by suspending your critical faculties make you accept more readily what the poet says.

A voice so thrilling ne'er was heard
In springtime from the cuckoo-bird,
Breaking the silence of the seas
Among the farthest Hebrides.

Matin Song

Pack, clouds, away! and welcome, day!
With night we banish sorrow.
Sweet air, blow soft; mount, lark, aloft
To give my Love good-morrow!
Wings from the wind to please her mind,
Notes from the lark I'll borrow:
Bird, prune thy wing! nightingale, sing!
To give my Love good-morrow!
To give my Love good-morrow
Notes from them all I'll borrow.

Wake from thy nest, robin red-breast!
Sing, birds, in every furrow!
And from each bill let music shrill
Give my fair Love good-morrow!

Blackbird and thrush, in every bush,
  Stare, linnet, and cock-sparrow,
You pretty elves, among yourselves
  Sing my fair Love good-morrow!
    To give my Love good-morrow
    Sing, birds, in every furrow!

THOMAS HEYWOOD

Stare] Starling.

Aubade

Hark! hark! the lark at heaven's gate sings,
  And Phoebus 'gins arise,
His steeds to water at those springs
  On chaliced flowers that lies;
And winking Mary-buds begin
  To ope their golden eyes:
With everything that pretty is,
  My lady sweet, arise!
    Arise, arise!

WILLIAM SHAKESPEARE

Hymn to Diana

Queen and huntress, chaste and fair,
Now the sun is laid to sleep,
Seated in thy silver chair,
State in wonted manner keep:
  Hesperus entreats thy light,
  Goddess, excellently bright.

Earth, let not thy envious shade
Dare itself to interpose;
Cynthia's shining orb was made
Heaven to clear, when day did close:
  Bless us then with wishèd sight,
  Goddess, excellently bright.

Lay thy bow of pearl apart,
And thy crystal-shining quiver;
Give unto the flying hart
Space to breathe, how short soever:
  Thou that mak'st a day of night,
  Goddess, excellently bright.

BEN JONSON

The Cherry-Tree Carol

As Joseph was a-walking,
  He heard an angel sing;
'This night shall be born
  Our Heavenly King.

'He neither shall be born
  In housen nor in hall,
Nor in the place of Paradise,
  But in an ox's stall.

'He neither shall be clothèd
  In purple nor in pall,
But all in fair linen,
  As were babies all.

'He neither shall be rock'd
  In silver nor in gold,
But in a wooden cradle
  That rocks on the mould.

'He neither shall be christen'd
  In white wine nor red
But with fair spring water,
  With which we were christenèd.'

At the Round Earth's Imagined Corners

At the round earth's imagined corners, blow
Your trumpets, angels, and arise, arise
From death, you numberless infinities
Of souls, and to your scattered bodies go;
All whom the flood did, and fire shall o'erthrow,
All whom war, dearth, age, agues, tyrannies,
Despair, law, chance hath slain, and you, whose eyes
Shall behold God, and never taste death's woe.
But let them sleep, Lord, and me mourn a space;
For, if above all these my sins abound,
'Tis late to ask abundance of Thy grace,
When we are there. Here on this lowly ground,
Teach me how to repent, for that's as good
As if thou hadst seal'd my pardon with thy blood.

JOHN DONNE

Song for St Cecilia's Day

(22 NOVEMBER 1687)

From harmony, from heavenly harmony
  This universal frame began.
When nature underneath a heap
  Of jarring atoms lay,
  And could not heave her head,
The tuneful voice was heard from high,
  Arise, ye more than dead.
Then cold, and hot, and moist, and dry,
In order to their stations leap,
  And Music's power obey.
From harmony, from heavenly harmony
  This universal frame began:
  From harmony to harmony
Through all the compass of the notes it ran,
The diapason closing full in Man.

What passion cannot Music raise and quell?
  When Jubal struck the corded shell,
His listening brethren stood around,
  And, wondering, on their faces fell
  To worship that celestial sound.
Less than a God they thought there could not dwell
  Within the hollow of that shell,
  That spoke so sweetly and so well.
What passion cannot Music raise and quell?

  The trumpet's loud clangour
    Excites us to arms,
  With shrill notes of anger;
    And mortal alarms.
  The double double double beat
      Of the thundering drum
      Cries, Hark! the foes come;
  Charge, charge, 'tis too late to retreat!

    The soft complaining flute
    In dying notes discovers
    The woes of hopeless lovers,
Whose dirge is whisper'd by the warbling lute.

    Sharp violins proclaim
  Their jealous pangs and desperation,
  Fury, frantic indignation,
  Depth of pains, and height of passion,
    For the fair disdainful dame.

    But O! what art can teach,
    What human voice can reach
      The sacred organ's praise?
    Notes inspiring holy love,
  Notes that wing their heavenly ways
    To mend the choirs above.

  Orpheus could lead the savage race,
  And trees unrooted left their place,
    Sequacious of the lyre;
But bright Cecilia rais'd the wonder higher:

When to her organ vocal breath was given,
  An angel heard, and straight appear'd,
    Mistaking earth for heaven!

GRAND CHORUS

As from the power of sacred lays
  The spheres began to move,
And sung the great Creator's praise
  To all the bless'd above;
So, when the last and dreadful hour
This crumbling pageant shall devour,
The trumpet shall be heard on high,
The dead shall live, the living die,
And Music shall untune the sky.

JOHN DRYDEN

Sweet Musick

Orpheus with his lute made trees,
And the mountain-tops, that freeze,
  Bow themselves, when he did sing:
To his musick, plants, and flowers,
Ever sprung; as sun, and showers,
  There had been a lasting spring.

Every thing that heard him play,
Even the billows of the sea,
  Hung their heads, and then lay by,
In sweet musick is such art;
Killing care, and grief of heart,
  Fall asleep, or, hearing, die.

JOHN FLETCHER

Everyone Sang

Everyone suddenly burst out singing;
And I was filled with such delight
As prisoned birds must find in freedom
Winging wildly across the white
Orchards and dark green fields; on; on; and out of sight.

Everyone's voice was suddenly lifted,
And beauty came like the setting sun.
My heart was shaken with tears and horror
Drifted away...O but everyone
Was a bird; and the song was wordless; the singing
 will never be done SIEGFRIED SASSOON

The Reaper

Behold her, single in the field,
Yon solitary Highland lass!
Reaping and singing by herself;
Stop here, or gently pass!
Alone she cuts and binds the grain,
And sings a melancholy strain;
Oh, listen! for the vale profound
Is overflowing with the sound.

No nightingale did ever chaunt
More welcome notes to weary bands
Of travellers in some shady haunt,
Among Arabian sands:
A voice so thrilling ne'er was heard
In spring-time from the cuckoo-bird,
Breaking the silence of the seas
Among the farthest Hebrides.

Will no one tell me what she sings?
Perhaps the plaintive numbers flow
For old, unhappy, far-off things,
And battles long ago:
Or is it some more humble lay,
Familiar matter of to-day?
Some natural sorrow, loss, or pain,
That has been, and may be again?

Whate'er the theme, the maiden sang
As if her song could have no ending;
I saw her singing at her work,
And o'er the sickle bending;
I listen'd, till I had my fill;
And, as I mounted up the hill,
The music in my heart I bore,
Long after it was heard no more.

WILLIAM WORDSWORTH

Mowing

There was never a sound beside the wood but one,
And that was my long scythe whispering to the ground.
What was it it whispered? I knew not well myself;
Perhaps it was something about the heat of the sun,
Something, perhaps, about the lack of sound—
And that was why it whispered and did not speak.
It was no dream of the gift of idle hours,
Or easy gold at the hand of fay or elf:
Anything more than the truth would have seemed too weak
To the earnest love that laid the swale in rows,
Not without feeble-pointed spikes of flowers
(Pale orchises), and scared a bright green snake.
The fact is the sweetest dream that labour knows.
My long scythe whispered and left the hay to make.

ROBERT FROST

Blow, Bugle, Blow

The splendour falls on castle walls
And snowy summits old in story:
The long light shakes across the lakes,
And the wild cataract leaps in glory.
Blow, bugle, blow, set the wild echoes flying,
Blow, bugle; answer, echoes, dying, dying, dying.

O hark, O hear! how thin and clear,
And thinner, clearer, farther going!
O sweet and fa from cliff and scar
The horns of Elfland faintly blowing!
Blow, let us hear the purple glens replying:
Blow, bugle; answer, echoes, dying, dying, dying.

O love, they die in yon rich sky,
They faint on hill or field or river:
Our echoes roll from soul to soul,
And grow for ever and for ever.
Blow, bugle, blow, set the wild echoes flying,
And answer, echoes, answer, dying, dying, dying.

LORD TENNYSON

The Lotos-Eaters: Choric Song

There is sweet music here that softer falls
Than petals from blown roses on the grass,
Or night-dews on still waters between walls
Of shadowy granite, in a gleaming pass;
Music that gentlier on the spirit lies,
Than tired eyelids upon tired eyes;
Music that brings sweet sleep down from the blissful skies.
Here are cool mosses deep,
And through the moss the ivies creep,
And in the stream the long-leaved flowers weep,
And from the craggy ledge the poppy hangs in sleep.

Why are we weighed upon with heaviness,
And utterly consumed with sharp distress,
While all things else have rest from weariness?
All things have rest: why should we toil alone,
We only toil, who are the first of things,
And make perpetual moan,
Still from one sorrow to another thrown:
Nor ever fold our wings,
And cease from wanderings,

Nor steep our brows in slumber's holy balm;
Nor hearken what the inner spirit sings,
'There is no joy but calm!'—
Why should we only toil, the roof and crown of things?

Lo! in the middle of the wood,
The folded leaf is wooed from out the bud
With winds upon the branch, and there
Grows green and broad, and takes no care,
Sun-steeped at noon, and in the moon
Nightly dew-fed; and turning yellow
Falls, and floats adown the air.
Lo! sweetened with the summer light,
The full-juiced apple, waxing over-mellow,
Drops in a silent autumn night.
All its allotted length of days,
The flower ripens in its place,
Ripens and fades, and falls, and hath no toil,
Fast-rooted in the fruitful soil.

Hateful is the dark-blue sky,
Vaulted o'er the dark-blue sea.
Death is the end of life; ah, why
Should life all labour be?
Let us alone. Time driveth onward fast,
And in a little while our lips are dumb.
Let us alone. What is it that will last?
All things are taken from us, and become
Portions and parcels of the dreadful Past.
Let us alone. What pleasure can we have
To war with evil? Is there any peace
In ever climbing up the climbing wave?
All things have rest, and ripen toward the grave
In silence; ripen, fall and cease:
Give us long rest or death, dark death, or dreamful ease.

How sweet it were, hearing the downward stream,
With half-shut eyes ever to seem
Falling asleep in a half-dream!

To dream and dream, like yonder amber light,
Which will not leave the myrrh-bush on the height;
To hear each other's whispered speech;
Eating the Lotos day by day,
To watch the crisping ripples on the beach,
And tender curving lines of creamy spray;
To lend our hearts and spirits wholly
To the influence of mild-minded melancholy;
To muse and brood and live again in memory,
With those old faces of our infancy
Heaped over with a mound of grass,
Two handfuls of white dust, shut in an urn of brass!

Dear is the memory of our wedded lives,
And dear the last embraces of our wives
And their warm tears: but all hath suffered change:
For surely now our household hearths are cold:
Our sons inherit us: our looks are strange:
And we should come like ghosts to trouble joy.
Or else the island princes over-bold
Have eat our substance, and the minstrel sings
Before them of the ten years' war in Troy,
And our great deeds, as half-forgotten things.
Is there confusion in the little isle?
Let what is broken so remain.
The Gods are hard to reconcile:
'Tis hard to settle order once again.
There *is* confusion worse than death,
Trouble on trouble, pain on pain,
Long labour unto agèd breath,
Sore task to hearts worn out with many wars
And eyes grown dim with gazing on the pilot-stars.

But, propt on beds of amaranth and moly,
How sweet (while warm airs lull us, blowing lowly)
With half-dropt eyelid still,
Beneath a heaven dark and holy,
To watch the long bright river drawing slowly
His waters from the purple hill—

To hear the dewy echoes calling
From cave to cave through the thick-twinèd vine—
To watch the emerald-coloured water falling
Through many a wov'n acanthus-wreath divine!
Only to hear and see the far-off sparkling brine,
Only to hear were sweet, stretched out beneath the pine.

The Lotos blooms below the barren peak:
The Lotos blows by every winding creek:
All day the wind breathes low with mellower tone:
Through every hollow cave and alley lone
Round and round the spicy downs the yellow Lotos-dust is blown.
We have had enough of action, and of motion we,
Rolled to starboard, rolled to larboard, when the surge was seething free,
Where the wallowing monster spouted his foam-fountains in the sea.
Let us swear an oath, and keep it with an equal mind,
In the hollow Lotos-land to live and lie reclined
On the hills like Gods together, careless of mankind.
For they lie beside their nectar, and the bolts are hurled
Far below them in the valleys, and the clouds are lightly curled
Round their golden houses, girdled with the gleaming world:
Where they smile in secret, looking over wasted lands,
Blight and famine, plague and earthquake, roaring deeps and fiery sands,
Clanging fights, and flaming towns, and sinking ships, and praying hands.
But they smile, they find a music centred in a doleful song
Steaming up, a lamentation and an ancient tale of wrong,
Like a tale of little meaning though the words are strong;
Chanted from an ill-used race of men that cleave the soil,
Sow the seed, and reap the harvest with enduring toil,
Storing yearly little dues of wheat, and wine and oil;
Till they perish and they suffer—some, 'tis whispered—down in hell

Suffer endless anguish, others in Elysian valleys dwell,
Resting weary limbs at last on beds of asphodel.
Surely, surely, slumber is more sweet than toil, the shore
Than labour in the deep mid-ocean, wind and wave and oar;
O rest ye, brother mariners, we will not wander more.

LORD TENNYSON

L'Allegro

Haste thee, Nymph, and bring with thee
Jest, and youthful Jollity,
Quips and Cranks and wanton Wiles,
Nods and Becks and wreathèd Smiles,
Such as hang on Hebe's cheek,
And love to live in dimple sleek;
Sport that wrinkled Care derides,
And Laughter holding both his sides.
Come, and trip it, as you go,
On the light fantastic toe;
And in thy right hand lead with thee
The mountain-nymph, sweet Liberty;
And, if I give thee honour due,
Mirth, admit me of thy crew,
To live with her, and live with thee,
In unreprovèd pleasures free;
To hear the lark begin his flight,
And, singing, startle the dull night,
From his watch-tower in the skies,
Till the dappled dawn doth rise;
Then to come, in spite of sorrow,
And at my window bid good-morrow,
Through the sweet-brier or the vine,
Or the twisted eglantine;
While the cock, with lively din,
Scatters the rear of darkness thin;
And to the stack, or the barn-door,
Stoutly struts his dames before:

Oft list'ning how the hounds and horn
Cheerly rouse the slumb'ring morn,
From the side of some hoar hill,
Through the high wood echoing shrill:
Sometime walking, not unseen,
By hedgerow elms, on hillocks green,
Right against the eastern gate
Where the great Sun begins his state,
Robed in flames and amber light,
The clouds in thousand liveries dight;
While the ploughman, near at hand,
Whistles o'er the furrow'd land,
And the milkmaid singeth blithe,
And the mower whets his scythe,
And every shepherd tells his tale
Under the hawthorn in the dale.
  Straight mine eye hath caught new pleasures,
Whilst the landskip round it measures:
Russet lawns, and fallows grey,
Where the nibbling flocks do stray;
Mountains on whose barren breast
The labouring clouds do often rest;
Meadows trim with daisies pied,
Shallow brooks, and rivers wide;
Towers and battlements it sees
Bosomed high in tufted trees,
Where perhaps some beauty lies,
The cynosure of neighbouring eyes.
  Hard by a cottage chimney smokes
From betwixt two aged oaks,
Where Corydon and Thyrsis met
Are at their savoury dinner set
Of herbs and other country messes,
Which the neat-handed Phillis dresses;
And then in haste her bower she leaves,
With Thestylis to bind the sheaves;
Or, if the earlier season lead,
To the tanned haycock in the mead.

Sometimes, with secure delight,
The upland hamlets will invite,
When the merry bells ring round,
And the jocund rebecks sound
To many a youth and many a maid
Dancing in the chequer'd shade,
And young and old come forth to play
On a sunshine holiday,
Till the livelong daylight fail:
Then to the spicy nut-brown ale,
With stories told of many a feat,
How Faery Mab the junkets eat.
She was pinch'd and pull'd, she said;
And he, by Friar's lantern led,
Tells how the drudging goblin sweat
To earn his cream-bowl duly set,
When in one night, ere glimpse of morn,
His shadowy flail hath thresh'd the corn
That ten day-labourers could not end;
Then lies him down, the lubber fiend,
And, stretch'd out all the chimney's length,
Basks at the fire his hairy strength;
And crop-full out of doors he flings,
Ere the first cock his matin rings.
Thus done the tales, to bed they creep,
By whispering winds soon lull'd asleep.
Tower'd cities please us then,
And the busy hum of men,
Where throngs of knights and barons bold,
In weeds of peace, high triumphs hold,
With store of ladies, whose bright eyes
Rain influence, and judge the prize
Of wit or arms, while both contend
To win her grace whom all commend.
There let Hymen oft appear
In saffron robe, with taper clear,
And pomp, and feast, and revelry,
With mask, and antique pageantry:

Such sights as youthful poets dream
On summer eves by haunted stream.
Then to the well-trod stage anon,
If Jonson's learned sock be on,
Or sweetest Shakespeare, Fancy's child,
Warble his native wood-notes wild.

And ever, against eating cares,
Lap me in soft Lydian airs,
Married to immortal verse,
Such as the meeting soul may pierce
In notes, with many a winding bout
Of linkèd sweetness long drawn out,
With wanton heed, and giddy cunning,
The melting voice through mazes running,
Untwisting all the chains that tie
The hidden soul of harmony;
That Orpheus' self may heave his head
From golden slumber on a bed
Of heap'd Elysian flowers, and hear
Such strains as would have won the ear
Of Pluto to have quite set free
His half-regain'd Eurydice.

These delights if thou canst give,
Mirth, with thee I mean to live.

JOHN MILTON

Fancy

Ever let the fancy roam,
Pleasure never is at home:
At a touch sweet Pleasure melteth,
Like to bubbles when rain pelteth;
Then let wingèd Fancy wander
Through the thought still spread beyond her:
Open wide the mind's cage-door,
She'll dart forth, and cloudward soar.

O sweet Fancy! let her loose;
Summer's joys are spoilt by use,
And the enjoying of the Spring
Fades as does its blossoming;
Autumn's red-lipp'd fruitage too,
Blushing through the mist and dew,
Cloys with tasting: What do then?
Sit thee by the ingle, when
The sear faggot blazes bright,
Spirit of a winter's night;
When the soundless earth is muffled,
And the cakèd snow is shuffled
From the ploughboy's heavy shoon;
When the Night doth meet the Noon
In a dark conspiracy
To banish Even from her sky.
Sit thee there, and send abroad,
With a mind self-overaw'd,
Fancy, high-commission'd:—send her!
She has vassals to attend her:
She will bring, in spite of frost,
Beauties that the earth hath lost;
She will bring thee, all together,
All delights of summer weather;
All the buds and bells of May,
From dewy sward or thorny spray;
All the heapèd Autumn's wealth,
With a still, mysterious stealth:
She will mix these pleasures up
Like three fit wines in a cup,
And thou shalt quaff it:—thou shalt hear
Distant harvest-carols clear;
Rustle of the reapèd corn;
Sweet birds antheming the morn:
And, in the same moment—hark!
'Tis the early April lark,
Or the rooks, with busy caw,
Foraging for sticks and straw.

Thou shalt, at one glance, behold
The daisy and the marigold;
White-plum'd lilies, and the first
Hedge-grown primrose that hath burst;
Shaded hyacinth, alway
Sapphire queen of the mid-May;
And every leaf, and every flower
Pearlèd with the self-same shower.
Thou shalt see the field-mouse peep
Meagre from its cellèd sleep;
And the snake all winter-thin
Cast on sunny bank its skin;
Freckled nest-eggs thou shalt see
Hatching in the hawthorn-tree,
When the hen-bird's wing doth rest
Quiet on her mossy nest;
Then the hurry and alarm
When the bee-hive casts its swarm;
Acorns ripe down-pattering,
While the autumn breezes sing.

Oh, sweet Fancy! let her loose;
Every thing is spoilt by use:
Where's the cheek that doth not fade,
Too much gaz'd at? Where's the maid
Whose lip mature is ever new?
Where's the eye, however blue,
Doth not weary? Where's the face
One would meet in every place?
Where's the voice, however soft,
One would hear so very oft?
At a touch sweet Pleasure melteth
Like to bubbles when rain pelteth.
Let, then, wingèd Fancy find
Thee a mistress to thy mind:
Dulcet-eyed as Ceres' daughter,
Ere the God of Torment taught her
How to frown and how to chide;
With a waist and with a side

White as Hebe's, when her zone
Slipt its golden clasp, and down
Fell her kirtle to her feet,
While she held the goblet sweet,
And Jove grew languid.—Break the mesh
Of the Fancy's silken leash;
Quickly break her prison-string
And such joys as these she'll bring.—
Let the wingèd Fancy roam,
Pleasure never is at home.

JOHN KEATS

SECTION 2

SCENES OF THE MACHINE AGE

A great deal of English poetry has been written describing the natural scene. The tradition of nature poetry goes back in time to ancient nature worship. But the world has changed so much since the industrial revolution that nature no longer plays a leading part in man's thoughts.

The poet, however, still reacts to the world around him, now changed almost beyond recognition from what it was before the discovery of steam and the invention of the machine. Instead of contemplating the solitude of mountains or the flight of the swallow, today he is excited by the tremendous power of a dynamo, by the thrilling speed of a jet-propelled aeroplane, or by the complicated hubbub of a great city.

He must think of fresh ways to describe these new scenes for the traditional way of expression, drawn from the natural scene, will no longer serve his purpose: he must experiment, invent, try out original techniques, making poetry from objects which may not appear to be material for poetry at all. We shall find new imagery used to describe the machine age, such as

the black statement of pistons

in the description of a train leaving the station. We shall be invited to imagine

the luminous self-possession of ships on ocean

and watch an air-liner:

More beautiful and soft than any moth
With burring furred antennae feeling its huge path
Through dusk.

New styles are created, vivid and arresting. Here is an express

shovelling white steam over her shoulder,

and a large modern city where

the cars' headlights bud,
Out from the sideroads, and the traffic signals, crême de menthe
or bulls' blood,
Tell one to stop, the engine gently breathing, or to go on
To where like black pipes of organs in the frayed and fading zone
Of the West the factory chimneys on sullen sentry will all night wait
To call, in the harsh morning, sleep-stupid faces through the daily
gate.

We find romance here also. Not the romance of

magic casements, opening on the foam
Of perilous seas in faery lands forlorn,

but of wireless and television, of sky-scrapers and telegraph pylons

along whose wires...
...far above and far as sight endures
Like whips of anger
With lightning's danger
There runs the quick perspective of the future.

Tugs

At noon three English dowagers ride
Stiff of neck and dignified,
Margaret, *Maud*, and *Mary Blake*,
With servile barges in their wake:

But silhouetted at midnight,
Darkly, by green and crimson light,
Three Nubian queens pass down the Thames
Statelily with flashing gems.

G. ROSTREVOR HAMILTON

Docks

When paint or steel or wood are wearing thin,
Then they come in:
The liners, schooners, merchantmen, and tramps,
Upon a head of water pressing hard
On gates of greenheart wood, that close and guard
The docks, till lintels, clamps,
Swing suddenly on quoins steel-pivoted,
With harsh complaint and clang,
And then above the walls arise and spread
Top-gallant yards or funnel, spanker-vang
Or dolphin-striker; figure-heads arise
That settling sway
Beside an inn; a mermaid's breasts and eyes
Beneath a bowsprit glare beside a dray.

All docks are wonderful, whether beside
The estuaries or foreshores robbed of sea,
Where jetties and much dredging keep them free,
And the strong constant scouring of the tide
Sweeps down the silt; or where by sandy dune
The neap-tides leave them dry, or flood-tides dash
With a vindictive lash
At the conjunction of the sun and moon.
And wonderful are dry docks, where the ships
Are run on keel-props held by timber-shores,
And sterns and prores
Stand up for scrapers' work, and the paint drips
Among algae and mussels; wonderful when
Docks still are in the building, and the pumps
Move water from the sumps,
And derricks, little trains, and shouting men
Dump clustered cylinders upon the gravel,
And through the sky square blocks of granite travel,

spanker-vang] rope holding a sail.
dolphin-striker] a short spar.
prores] prows. algae] seaweed.

Dangling to place to make the sills. Or when
As now by Thames the running currents flush
The sluices of the locks, and seek to rush
Reverse-gate strengthening the entrances,
Harry the boats, and shift
The refuse of the town and littoral drift;
And in the dusk the slums are palaces.
They wait upon the sea.
And wharf and jetty, stately in the grime
Make commerce classical, and turn sublime
The warehouse crammed with jute or flax or tea.

DOROTHY WELLESLEY

Journey

I

How many times I nearly miss the train
By running up the staircase once again
For some dear trifle almost left behind.
At that last moment the unwary mind
Forgets the solemn tick of station-time;
That muddy lane the feet must climb—
The bridge—the ticket—signal down—
Train just emerging beyond the town:
The great blue engine panting as it takes
The final curve, and grinding on its brakes
Up to the platform-edge.... The little doors
Swing open, while the burly porter roars.
The tight compartment fills: our careful eyes
Go to explore each other's destinies.
A lull. The station-master waves. The train
Gathers, and grips, and takes the rails again,
Moves to the shining open land, and soon
Begins to tittle-tattle a tame tattoon.

II

They ramble through the country-side,
Dear gentle monsters, and we ride
Pleasantly seated—so we sink
Into a torpor on the brink
Of thought, or read our books, and understand
Half them and half the backward-gliding land:
(Trees in a dance all twirling round;
Large rivers flowing with no sound;
The scattered images of town and field,
Shining flowers half concealed.)
And, having settled to an equal rate,
They swing the curve and straighten to the straight,
Curtail their stride and gather up their joints,
Snort, dwindle their steam for the noisy points,
Leap them in safety, and, the other side,
Loop again to an even stride.

The long train moves: we move in it along.
Like an old ballad, or an endless song,
It drones and wimbles its unwearied croon—
Croons, drones, and mumbles all the afternoon.

Towns with their fifty chimneys close and high,
Wreathed in great smoke between the earth and sky,
It hurtles through them, and you think it must
Halt—but it shrieks and sputters them with dust,
Cracks like a bullet through their big affairs,
Rushes the station-bridge, and disappears
Out to the suburb, laying bare
Each garden trimmed with pitiful care;
Children are caught at idle play,
Held a moment, and thrown away.
Nearly everyone looks round.
Some dignified inhabitant is found
Right in the middle of the commonplace—
Buttoning his trousers, or washing his face.

III

Oh, the wild engine! Every time I sit
In any train I must remember it.
The way it smashes through the air; its great
Petulant majesty and terrible rate:
Driving the ground before it, with those round
Feet pounding, eating, covering the ground;
The piston using up the white steam so
You cannot watch it when it come or go;
The cutting, the embankment; how it takes
The tunnels, and the clatter that it makes;
So careful of the train and of the track,
Guiding us out, or helping us go back;
Breasting its destination: at the close
Yawning, and slowly dropping to a doze.

IV

We who have looked each other in the eyes
This journey long, and trundled with the train,
Now to our separate purposes must rise,
Becoming decent strangers once again.
The little chamber we have made our home
In which we so conveniently abode,
The complicated journey we have come,
Must be an unremembered episode.
Our common purpose made us all like friends.
How suddenly it ends!
A nod, a murmur, or a little smile,
Or often nothing, and away we file.
I hate to leave you, comrades. I will stay
To watch you drift apart and pass away.
It seems impossible to go and meet
All those strange eyes of people in the street.
But, like some proud unconscious god, the train
Gathers us up and scatters us again.

HAROLD MONRO

In the Train

I am in a long train gliding through England,
Gliding past green fields and gentle grey willows,
Past huge dark elms and meadows full of buttercups,
And old farms dreaming among mossy apple trees.

Now we are in a dingy town of small ugly houses
And tin advertisements of cocoa and Sunlight Soap,
Now we are in dreary station built of coffee-coloured wood,
Where barmaids in black stand in empty Refreshment Rooms,
And shabby old women sit on benches with suitcases.

Now we are by sidings where coaltrucks lurk disconsolate
Bright skies overarch us with shining cloud palaces,
Sunshine flashes on canals, and then the rain comes,
Silver rain from grey skies lashing our window panes;
Then it is bright again and white smoke is blowing
Gaily over a pale blue sky among the telegraph wires.

Northward we rush under bridges, up gradients,
Through black, smoky tunnels, over iron viaducts,
Past platelayers and signal boxes, factories and warehouses;
Afternoon is fading among the tall brick chimney-stacks
In the murky Midlands where meadows grow more colourless.
Northward, O train, you rush, resolute, invincible,
Northward to the night where your banner of flying smoke
Will glow in the darkness with burning spark and ruddy flame.

Be the train, my life, see the shining meadows,
Glance at the quiet farms, the gardens and shady lanes,
But do not linger by them, look at the dingy misery
Of all those silly towns, see it, hate it and remember it,
But never accept it. You must only accept your own road:
The strong unchanging steel rails of necessity,
The ardent power that drives you towards night and the
unknown terminus.

V. DE SOLA PINTO

Morning Express

Along the wind-swept platform, pinched and white,
The travellers stand in pools of wintry light,
Offering themselves to morn's long slanting arrows.
The train's due; porters trundle laden barrows.
The train steams in, volleying resplendent clouds
Of sun-blown vapour. Hither and about,
Scared people hurry, storming the doors in crowds.
The officials seem to waken with a shout,
Resolved to hoist and plunder; some to the vans
Leap; others rumble the milk in gleaming cans.

Boys, indolent-eyed, from baskets leaning back,
Question each face; a man with a hammer steals
Stooping from coach to coach; with clang and clack,
Touches and tests, and listens to the wheels.
Guard sounds a warning whistle, points to the clock
With brandished flag, and on his folded flock
Claps the last door: the monster grunts: 'Enough!'
Tightening his load of links with pant and puff.
Under the arch, then forth into blue day;
Glide the processional windows on their way,
And glimpse the stately folk who sit at ease
To view the world like kings taking the seas
In prosperous weather: drifting banners tell
Their progress to the counties; with them goes
The clamour of their journeying; while those
Who sped them stand to wave a last farewell.

SIEGFRIED SASSOON

The Bridge

Here, with one leap,
The bridge that spans the cutting; on its back
The load
Of the main-road,
And under it the railway-track.

Into the plains they sweep,
Into the solitary plains asleep,
The flowing lines, the parallel lines of steel—
Fringed with their narrow grass,
Into the plains they pass,
The flowing lines, like arms of mute appeal.

A cry
Prolonged across the earth—a call
To the remote horizons and the sky;
The whole east rushes down them with its light,
And the whole west receives them, with its pall
Of stars and night—
The flowing lines, the parallel lines of steel.

And with the fall
Of darkness, see! the red,
Bright anger of the signal, where it flares
Like a huge eye that stares
On some hid danger in the dark ahead.
A twang of wire—unseen
The signal drops; and now, instead
Of a red eye, a green.

Out of the silence grows
An iron thunder—grows, and roars, and sweeps,
Menacing! The plain
Suddenly leaps,
Startled, from its repose—
Alert and listening. Now, from the gloom
Of the soft distance, loom
Three lights and, over them, a brush
Of tawny flame and flying spark—
Three pointed lights that rush,
Monstrous, upon the cringing dark.

And nearer, nearer rolls the sound,
Louder the throb and roar of wheels,
The shout of speed, the shriek of steam;
The sloping bank,

Cut into flashing squares, gives back the clank
And grind of metal, while the ground
Shudders and the bridge reels—
As, with a scream,
The train,
A rage of smoke, a laugh of fire,
A lighted anguish of desire,
A dream
Of gold and iron, of sound and flight,
Tumultuous roars across the night.

The train roars past—and, with a cry,
Drowned in a flying howl of wind,
Half-stifled in the smoke and blind,
The plain,
Shaken, exultant, unconfined,
Rises, flows on, and follows, and sweeps by,
Shrieking, to lose itself in distance and the sky.

JOHN REDWOOD ANDERSON

The Express

After the first powerful plain manifesto
The black statement of pistons, without more fuss
But gliding like a queen, she leaves the station.
Without bowing and with restrained unconcern
She passes the houses which humbly crowd outside,
The gasworks and at last the heavy page
Of death, printed by gravestones in the cemetery.
Beyond the town there lies the open country
Where, gathering speed, she acquires mystery,
The luminous self-possession of ships on ocean.
It is now she begins to sing—at first quite low
Then loud, and at last with a jazzy madness—
The song of her whistle screaming at curves,
Of deafening tunnels, brakes, innumerable bolts.

And always light, aerial, underneath
Goes the elate metre of her wheels.
Steaming through metal landscape on her lines
She plunges new eras of wild happiness
Where speed throws up strange shapes, broad curves
And parallels clean like the steel of guns.
At last, further than Edinburgh or Rome,
Beyond the crest of the world, she reaches night
Where only a low streamline brightness
Of phosphorus on the tossing hills is white.
Ah, like a comet through flame she moves entranced
Wrapt in her music no bird song, no, nor bough
Breaking with honey buds, shall ever equal.

STEPHEN SPENDER

Night Mail

This is the night mail crossing the border,
Bringing the cheque and the postal order,
Letters for the rich, letters for the poor,
The shop at the corner and the girl next door.
Pulling up Beattock, a steady climb—
The gradient's against her, but she's on time.

Past cotton grass and moorland boulder
Shovelling white steam over her shoulder,
Snorting noisily as she passes
Silent miles of wind-bent grasses.
Birds turn their heads as she approaches,
Stare from the bushes at her blank-faced coaches.
Sheep dogs cannot turn her course,
They slumber on with paws across.
In the farm she passes no one wakes,
But a jug in the bedroom gently shakes.

Dawn freshens, the climb is done.
Down towards Glasgow she descends
Towards the steam tugs yelping down the glade of cranes.

Towards the fields of apparatus, the furnaces
Set on the dark plain like gigantic chessmen.
All Scotland waits for her:
In the dark glens, beside the pale-green lochs
Men long for news.

Letters of thanks, letters from banks,
Letters of joy from girl and boy,
Receipted bills and invitations
To inspect new stock or visit relations,
And applications for situations
And timid lovers' declarations
And gossip, gossip from all the nations,
News circumstantial, news financial,
Letters with holiday snaps to enlarge in,
Letters with faces scrawled in the margin,
Letters from uncles, cousins and aunts,
Letters to Scotland from the South of France,
Letters of condolence to Highlands and Lowlands,
Notes from overseas to Hebrides—
Written on paper of every hue,
The pink, the violet, the white and the blue,
The chatty, the catty, the boring, adoring,
The cold and official and the heart's outpouring,
Clever, stupid, short and long,
The typed and the printed and the spelt all wrong.

Thousands are still asleep
Dreaming of terrifying monsters,
Or a friendly tea beside the band at Cranston's or Crawford's:
Asleep in working Glasgow, asleep in well-set Edinburgh,
Asleep in granite Aberdeen.
They continue their dreams;
But shall wake soon and long for letters,
And none will hear the postman's knock
Without a quickening of the heart,
For who can hear and feel himself forgotten?

W. H. AUDEN

London to Paris, by Air

I

The droning roar is quickened, and we lift
On steady wing, like upward sweep of air,
Into the fleece-strewn heaven. The great plane
Draws to herself the leagues: onward we bear
In one resistless eddy towards the south
Over the English fields, trim-hedged and square,
And countless, winding lanes, a vast expanse
Of flattened green: huge shapes of shadow float
Inconsequent as bubbles: haunts of men
Stripped of their cherished privacy we note
And crawling multitudes within a town—
On all we rangers of the wind look down.

II

The coast-line swings to us: beneath our feet
The gray-green carpet of the sliding sea
Stretches afar, on it small, busy ships
Whose comet-tails in foamy whiteness flee:
We lift, and snowy cloudlets roam below,
Frail, wistful spirits of pure charity
Blessing the waters: like green marble veined,
The waves roll in upon the yellowing sand,
Then break to myriad, filmy curves of lace
Where they eternally caress the land:
Now low lies France—the kingdom of the breeze
Parts not the nations like the severing seas.

III

Down the wide river, jauntily outspread,
A fishing fleet comes seaward, to our eyes
Mere walnut shells with autumn leaves for sails:
And now a fellow pilgrim of the skies,
Like a big insect droning past our flank,
Cruises to England home: before us lies

The rolling plain with its great, hedgeless strips
Of close-tilled fields, red roofs, and pointed trees,
The feathered arrows of the long French roads,
And all the stretch of quiet harmonies:
Then haven shows, and downward to earth's breast,
Like homing bird, we wheel and sink to rest.

LORD GORELL

The Landscape near an Aerodrome

More beautiful and soft than any moth
With burring furred antennae feeling its huge path
Through dusk, the air-liner with shut-off engines
Glides over suburbs and the sleeves set trailing tall
To point the wind. Gently, broadly, she falls
Scarcely disturbing charted currents of air.

Lulled by descent, the travellers across sea
And across feminine land indulging its easy limbs
In miles of softness, now let their eyes trained by watching
Penetrate through dusk the outskirts of this town
Here where industry shows a fraying edge.
Here they may see what is being done.

Beyond the winking masthead light
And the landing-ground, they observe the outposts
Of work: chimneys like lank black fingers
Or figures frightening and mad: the squat buildings
With their strange air behind trees, like women's faces
Shattered by grief. Here where few houses
Moan with faint light behind their blinds
They remark the unhomely sense of complaint, like a dog
Shut out and shivering at the foreign moon.

In the last sweep of love, they pass over fields
Behind the aerodrome, where boys play all day
Hacking dead grass: whose cries, like wild birds,
Settle upon the nearest roofs
But soon are hid under the loud city.

Then, as they land, they hear the tolling bell
Reaching across the landscape of hysteria
To where, larger than all the charcoaled batteries
And imaged towers against that dying sky,
Religion stands, the church blocking the sun.

STEPHEN SPENDER

The Pylons

The secret of these hills was stone, and cottages
Of that stone made,
And crumbling roads
That turned on sudden hidden villages.

Now over these small hills they have built the concrete
That trails black wire:
Pylons, those pillars
Bare like nude, giant girls that have no secret.

The valley with its gilt and evening look
And the green chestnut
Of customary root
Are mocked dry like the parched bed of a brook.

But far above and far as sight endures
Like whips of anger
With lightning's danger
There runs the quick perspective of the future.

This dwarfs our emerald country by its trek
So tall with prophecy:
Dreaming of cities
Where often clouds shall lean their swan-white neck.

STEPHEN SPENDER

Crucifixion of the Skyscraper

Men took the skyscraper
And nailed it to the rock. Each nerve and vein
Were searched by iron hammers. Hour on hour,
The bolts were riveted tighter. Steel and stone
Did what they could to quench the fiery core
That blazed within. Till when the work was done,
Solid as a sepulchre, square-rooted to the rock,
The skyscraper, a well-polished tomb of hope,
Guarded by busy throngs of acolytes,
Shouldered aside the sun. Within its walls
Men laid a little gold.
But yet not dead
However long battered by furious life,
However buried under tons of frozen weight
That structure was. At night when crowds no more
Jostled its angles, but the weary streets
Of a worn planet stared out at the stars;
Its towering strength grown ghostly, pure, remote,
Lone on the velvety night in flights of gold
The tower rose. The skyscraper dripped light.

J. GOULD FLETCHER

SECTION 3

STORIES OF PURE IMAGINATION

Often in moments of relaxation our thoughts begin to wander, one thought leads to the next, one memory recalls another until we find ourselves a long way from our starting-point. We may go back along the chain of thoughts and discover the varied associations that led one to another. Long-forgotten memories may have been revived in this process of free association; it is free because it is not consciously controlled by any purpose, nor directed towards any goal.

When we indulge in day-dreams, building castles in the air, we give a shape and continuity to our thoughts, imagining things as we would like them to be rather than as they are in real life. In a day-dream we make up a story with ourselves as hero, having the most wonderful adventures, outdoing every rival, overcoming incredible difficulties and winning the most satisfying rewards, a formula very common on the films. Sometimes we imagine ourselves as a suffering hero, misunderstood by our friends, forsaken and unhappy, enlarging upon our failures in real life so that we become more impressive in relation to our woes and so regain our sense of self-importance.

The motives of the day-dream are very much the same as those of daily life, a desire to be successful; and the stuff of the dreams comes from the experiences of our daily life. It is obvious that the imagination plays a great part here, creating stories similar to the dream stories a poet puts into verse. For the poet lives in the world of his imagination:

We are the music makers,
  And we are the dreamers of dreams,

Wandering by lone sea-breakers,
And sitting by desolate streams.

Such poems will be romantic, dealing with tales and legends of the past, distant in place as well as in time; and they will be shaped by the rich fantasies of a wandering imagination:

Under tower and balcony,
By garden-wall and gallery,
A gleaming shape she floated by
Dead-pale between the houses high,
Silent into Camelot.

The poet may be the hero, or the victim, in the story he tells:

And there she lullèd me asleep
And there I dream'd—Ah! woe betide,
The latest dream I ever dream'd
On the cold hill side.

Such stories of the imagination have the various elements of poetry strongly marked. They have a pleasing rhythm, which lulls the critical reasoning part of the mind asleep gaining a 'willing suspension of our disbelief' while it excites and stimulates the emotions; they present a series of vivid images appealing to all the senses, and they tell a story.

These poems range from heaven to earth and from earth to heaven, embarking the reader upon a voyage as enthralling as that of the ancient Mariner.

The Dirge of Lovely Rosabelle

O listen, listen, ladies gay!
No haughty feat of arms I tell;
Soft is the note, and sad the lay,
That mourns the lovely Rosabelle.

'Moor, moor the barge, ye gallant crew!
And, gentle ladye, deign to stay!
Rest thee in Castle Ravensheuch,
Nor tempt the stormy firth to-day.

'The blackening wave is edged with white:
To inch and rock the sea-mews fly;
The fishers have heard the Water-Sprite,
Whose screams forebode that wreck is nigh.

'Last night the gifted Seer did view
A wet shroud swathed round ladye gay;
Then stay thee, Fair, in Ravensheuch:
Why cross the gloomy firth to-day?'

''Tis not because Lord Lindesay's heir
To-night at Roslin leads the ball,
But that my ladye-mother there
Sits lonely in her castle-hall.

''Tis not because the ring they ride,
And Lindesay at the ring rides well,
But that my sire the wine will chide,
If 'tis not filled by Rosabelle.'

O'er Roslin all that dreary night
A wondrous blaze was seen to gleam;
'Twas broader than the watch-fire's light,
And redder than the bright moonbeam.

It glared on Roslin's castled rock,
It ruddied all the copse-wood glen;
'Twas seen from Dryden's groves of oak,
And seen from caverned Hawthornden.

Seemed all on fire that chapel proud,
Where Roslin's chiefs uncoffined lie,
Each Baron for a sable shroud,
Sheathed in his iron panoply.

inch] isle.

Seemed all on fire within, around,
  Deep sacristy and altar's pale;
Shone every pillar foliage-bound,
  And glimmered all the dead men's mail.

Blazed battlement and pinnet high,
  Blazed every rose-carved buttress fair—
So still they blaze when fate is nigh
  The lordly line of high St Clair.

There are twenty of Roslin's barons bold
  Lie buried within that proud chapelle;
Each one the holy vault doth hold—
  But the sea holds lovely Rosabelle!

And each St Clair was buried there,
  With candle, with book, and with knell;
But the sea-caves rung, and the wild winds sung,
  The dirge of lovely Rosabelle.

SIR WALTER SCOTT

The Rime of the Ancient Mariner

(IN SEVEN PARTS)

PART I

It is an ancient Mariner,
And he stoppeth one of three.
'By thy long grey beard and glittering eye,
Now wherefore stopp'st thou me?

An ancient Mariner meeteth three Gallants bidden to a wedding-feast, and detaineth one.

The Bridegroom's doors are opened wide,
And I am next of kin;
The guests are met, the feast is set:
May'st hear the merry din.'

He holds him with his skinny hand,
'There was a ship,' quoth he.
'Hold off! unhand me, grey-beard loon!'
Eftsoons his hand dropt he.

The Wedding-Guest is spell-bound by the eye of the old seafaring man and constrained to hear his tale.

He holds him with his glittering eye—
The Wedding-Guest stood still,
And listens like a three years' child:
The Mariner hath his will.

The Wedding-Guest sat on a stone:
He cannot choose but hear;
And thus spake on that ancient man,
The bright-eyed Mariner.

'The ship was cheered, the harbour cleared,
Merrily did we drop
Below the kirk, below the hill,
Below the lighthouse top.

The Mariner tells how the ship sailed southward with a good wind and fair weather, till it reached the Line.

The Sun came up upon the left,
Out of the sea came he!
And he shone bright, and on the right
Went down into the sea.

Higher and higher every day,
Till over the mast at noon—'
The Wedding-Guest here beat his breast,
For he heard the loud bassoon.

The Wedding-Guest heareth the bridal music; but the Mariner continueth his tale.

The bride hath paced into the hall,
Red as a rose is she;
Nodding their heads before her goes
The merry minstrelsy.

The Wedding-Guest he beat his breast,
Yet he cannot choose but hear;
And thus spake on that ancient man,
The bright-eyed Mariner.

The ship driven by a storm toward the South Pole.

'And now the Storm-blast came, and he
Was tyrannous and strong;
He struck with his o'ertaking wings,
And chased us south along.

With sloping masts and dipping prow,
As who pursued with yell and blow
Still treads the shadow of his foe,
And forward bends his head,
The ship drove fast, loud roared the blast,
And southward aye we fled.

And now there came both mist and snow,
And it grew wondrous cold:
And ice, mast-high, came floating by,
As green as emerald.

And through the drifts the snowy clifts
Did send a dismal sheen:
Nor shapes of men nor beasts we ken—
The ice was all between.

The land of ice, and of fearful sounds where no living thing was to be seen.

The ice was here, the ice was there,
The ice was all around:
It cracked and growled, and roared and howled,
Like noises in a swound!

At length did cross an Albatross,
Thorough the fog it came;
As if it had been a Christian soul,
We hailed it in God's name.

Till a great sea-bird, called the Albatross, came through the snow-fog and was received with great joy and hospitality.

It ate the food it ne'er had eat,
And round and round it flew.
The ice did split with a thunder-fit;
The helmsman steered us through!

And a good south wind sprung up behind;
The Albatross did follow,
And every day, for food or play,
Came to the mariner's hollo!

And lo! the Albatross proveth a bird of good omen, and followeth the ship as it returned northward through fog and floating ice.

In mist or cloud, on mast or shroud,
It perched for vespers nine;
Whiles all the night, through fog-smoke white.
Glimmered the white moonshine.'

The ancient Mariner inhospitably killeth the pious bird of good omen.

'God save thee, ancient Mariner!
From the fiends, that plague thee thus!—
Why look'st thou so?'—'With my cross-bow
I shot the Albatross.

PART II

The Sun now rose upon the right:
Out of the sea came he,
Still hid in mist, and on the left
Went down into the sea.

And the good south wind still blew behind,
But no sweet bird did follow,
Nor any day for food or play
Came to the mariner's hollo!

His shipmates cry out against the ancient Mariner, for killing the bird of good luck.

And I had done a hellish thing,
And it would work 'em woe:
For all averred, I had killed the bird
That made the breeze to blow.
Ah, wretch! said they, the bird to slay,
That made the breeze to blow!

But when the fog cleared off, they justify the same, and thus make themselves accomplices in the crime.

Nor dim nor red, like God's own head,
The glorious Sun uprist:
Then all averred, I had killed the bird
That brought the fog and mist.
'Twas right, said they, such birds to slay,
That bring the fog and mist.

The fair breeze continues; the ship enters the Pacific Ocean, and sails northward, even till it reaches the Line.

The fair breeze blew, the white foam flew,
The furrow followed free;
We were the first that ever burst
Into that silent sea.

The ship hath been suddenly becalmed.

Down dropt the breeze, the sails dropt down,
'Twas sad as sad could be;
And we did speak only to break
The silence of the sea!

All in a hot and copper sky,
The bloody Sun, at noon,
Right up above the mast did stand,
No bigger than the Moon.

Day after day, day after day,
We stuck, nor breath nor motion;
As idle as a painted ship
Upon a painted ocean.

Water, water, everywhere,
And all the boards did shrink;
Water, water, everywhere
Nor any drop to drink.

And the Albatross begins to be avenged.

The very deep did rot: O Christ!
That ever this should be!
Yea, slimy things did crawl with legs
Upon the slimy sea.

About, about, in reel and rout
The death-fires danced at night;
The water, like a witch's oils,
Burnt green, and blue and white.

And some in dreams assurèd were
Of the Spirit that plagued us so,
Nine fathom deep he had followed us
From the land of mist and snow.

A Spirit had followed them; one of the invisible inhabitants of this planet, neither departed souls nor angels; concerning whom the learned Jew, Josephus, and the Platonic Constantinopolitan Michael Psellus, may be consulted. They are very numerous, and there is no climate or element without one or more.

And every tongue, through utter drought,
Was withered at the root;
We could not speak, no more than if
We had been choked with soot.

Ah! well a-day! what evil looks
Had I from old and young!
Instead of the cross, the Albatross
About my neck was hung.

The shipmates, in their sore distress, would fain throw the whole guilt on the ancient Mariner: in sign whereof they hang the dead sea-bird round his neck.

PART III

There passed a weary time. Each throat
Was parched, and glazed each eye.
A weary time! a weary time!
How glazed each weary eye!
When looking westward, I beheld
A something in the sky.

The ancient Mariner beholdeth a sign in the element afar off.

At first it seemed a little speck,
And then it seemed a mist;
It moved and moved, and took at last
A certain shape, I wist.

A speck, a mist, a shape, I wist!
And still it neared and neared:
As if it dodged a water-sprite,
It plunged and tacked and veered.

At its nearer approach it seemeth him to be a ship; and at a dear ransom he freeth his speech from the bonds of thirst.

With throats unslaked, with black lips baked,
We could nor laugh nor wail;
Through utter drought all dumb we stood!
I bit my arm, I sucked the blood,
And cried, A sail! a sail!

A flash of joy;

With throats unslaked, with black lips baked,
Agape they heard me call;
Gramercy! they for joy did grin,
And all at once their breath drew in,
As they were drinking all.

And horror follows. For can it be a ship that comes onward without wind or tide?

See! see! (I cried) she tacks no more!
Hither to work us weal;
Without a breeze, without a tide,
She steadies with upright keel!

The western wave was all a-flame.
The day was well nigh done!
Almost upon the western wave
Rested the broad, bright Sun;
When that strange shape drove suddenly
Betwixt us and the Sun.

And straight the Sun was flecked with bars,
(Heaven's Mother send us grace!)
As if through a dungeon-grate he peered
With broad and burning face.

It seemeth him but the skeleton of a ship.

Alas! (thought I, and my heart beat loud)
How fast she nears and nears!
Are those her sails that glance in the Sun,
Like restless gossameres?

Are those her ribs through which the Sun
Did peer, as through a grate?
And is that Woman all her crew?
Is that a Death? and are there two?
Is Death that woman's mate?

And its ribs are seen as bars on the face of the setting Sun. The Spectre-Woman and her Death-mate, and no other on board the skeleton-ship.

Her lips were red, her looks were free,
Her locks were yellow as gold:
Her skin was as white as leprosy,
The Night-mare Life-in-Death was she,
Who thicks man's blood with cold.

Like vessel, like crew!

The naked hulk alongside came,
And the twain were casting dice;
"The game is done! I've won! I've won!"
Quoth she, and whistles thrice.

Death and Life-in-Death have diced for the ship's crew and she (the latter) winneth the ancient Mariner.

The Sun's rim dips; the stars rush out:
At one stride comes the dark;
With far-heard whisper, o'er the sea,
Off shot the spectre-bark.

No twilight within the courts of the Sun.

At the rising of the Moon,

We listened and looked sideways up!
Fear at my heart, as at a cup,
My life-blood seemed to sip!
The stars were dim, and thick the night,
The steersman's face by his lamp gleamed white;
From the sails the dew did drip—
Till clomb above the eastern bar
The hornèd Moon, with one bright star
Within the nether tip.

One after another,

One after one, by the star-dogged Moon,
Too quick for groan or sigh,
Each turned his face with a ghastly pang,
And cursed me with his eye.

His shipmates drop down dead.

Four times fifty living men,
(And I heard nor sigh nor groan)
With heavy thump, a lifeless lump,
They dropped down one by one.

But Life-in-Death begins her work on the ancient Mariner.

The souls did from their bodies fly,—
They fled to bliss or woe!
And every soul, it passed me by,
Like the whizz of my crossbow!'

PART IV

The Wedding-Guest feareth that a Spirit is talking to him.

'I fear thee, ancient Mariner!
I fear thy skinny hand!
And thou art long, and lank, and brown,
As is the ribbed sea-sand.

But the ancient Mariner assureth him of his bodily life, and proceedeth to relate his horrible penance.

I fear thee and thy glittering eye,
And thy skinny hand, so brown.'—
'Fear not, fear not, thou Wedding-Guest!
This body dropt not down.

Alone, alone, all all alone,
Alone on a wide wide sea!
And never a saint took pity on
My soul in agony.

The many men, so beautiful!
And they all dead did lie:
And a thousand thousand slimy things
Lived on; and so did I.

He despiseth the creatures of the calm.

I looked upon the rotting sea,
And drew my eyes away;
I looked upon the rotting deck,
And there the dead men lay.

And envieth that they should live, and so many lie dead.

I looked to heaven, and tried to pray;
But or ever a prayer had gusht,
A wicked whisper came, and made
My heart as dry as dust.

I closed my lids, and kept them close,
And the balls like pulses beat;
For the sky and the sea, and the sea and the sky
Lay like a load on my weary eye,
And the dead were at my feet.

The cold sweat melted from their limbs,
Nor rot nor reek did they:
The look with which they looked on me
Had never passed away.

But the curse liveth for him in the eye of the dead men.

An orphan's curse would drag to hell
A spirit from on high;
But oh! more horrible than that
Is the curse in a dead man's eye!
Seven days, seven nights, I saw that curse,
And yet I could not die.

In his loneliness and fixedness he yearneth towards the journeying Moon, and the stars that still sojourn, yet still move onward; and everywhere the blue sky belongs to them, and is their appointed rest, and their native country and their own natural homes which they enter unannounced, as lords that are certainly expected and yet there is a silent joy at their arrival.

The moving Moon went up the sky,
And nowhere did abide:
Softly she was going up,
And a star or two beside—

Her beams bemocked the sultry main,
Like April hoar-frost spread;
But where the ship's huge shadow lay,
The charmèd water burnt alway
A still and awful red.

By the light of the Moon he beholdeth God's creatures of the great calm.

Beyond the shadow of the ship,
I watched the water-snakes:
They moved in tracks of shining white,
And when they reared, the elfish light
Fell off in hoary flakes.

Within the shadow of the ship
I watched their rich attire:
Blue, glossy green, and velvet black,
They coiled and swam; and every track
Was a flash of golden fire.

Their beauty and their happiness.

O happy living things! no tongue
Their beauty might declare:
A spring of love gushed from my heart,
And I blessed them unaware:

He blesseth them in his heart.

Sure my kind saint took pity on me,
And I blessed them unaware.

The spell begins to break.

The selfsame moment I could pray;
And from my neck so free
The Albatross fell off, and sank
Like lead into the sea.

PART V

Oh sleep! it is a gentle thing,
Beloved from pole to pole!
To Mary Queen the praise be given!
She sent the gentle sleep from Heaven,
That slid into my soul.

The silly buckets on the deck,
That had so long remained,
I dreamt that they were filled with dew;
And when I awoke, it rained.

By grace of the holy Mother, the ancient Mariner is refreshed with rain.

My lips were wet, my throat was cold,
My garments all were dank;
Sure I had drunken in my dreams,
And still my body drank.

I moved, and could not feel my limbs;
I was so light—almost
I thought that I had died in sleep,
And was a blessèd ghost.

And soon I heard a roaring wind:
It did not come anear;
But with its sound it shook the sails,
That were so thin and sere.

He heareth sounds and seeth strange sights and commotions in the sky and the element.

The upper air burst into life!
And a hundred fire-flags sheen,
To and fro they were hurried about!
And to and fro, and in and out,
The wan stars danced between.

And the coming wind did roar more loud,
And the sails did sigh like sedge;
And the rain poured down from one black cloud;
The Moon was at its edge.

The thick black cloud was cleft, and still
The Moon was at its side:
Like waters shot from some high crag,
The lightning fell with never a jag,
A river steep and wide.

The bodies of the ship's crew are inspired, and the ship moves on;

The loud wind never reached the ship,
Yet now the ship moved on!
Beneath the lightning and the Moon
The dead men gave a groan.

They groaned, they stirred, they all uprose,
Nor spake, nor moved their eyes;
It had been strange, even in a dream,
To have seen those dead men rise.

The helmsman steered, the ship moved on;
Yet never a breeze up blew;
The mariners all 'gan work the ropes,
Where they were wont to do;
They raised their limbs like lifeless tools—
We were a ghastly crew.

The body of my brother's son
Stood by me, knee to knee:
The body and I pulled at one rope,
But he said nought to me.'

But not by the souls of the men nor by demons of earth or middle air, but by a blessed troop of angelic spirits, sent down by the invocation of the guardian saint.

'I fear thee, ancient Mariner!'
'Be calm, thou Wedding-Guest!
'Twas not those souls that fled in pain,
Which to their corses came again,
But a troop of spirits blest:

For when it dawned—they dropped their arms,
And clustered round the mast;
Sweet sounds rose slowly through their mouths,
And from their bodies passed.

Around, around, flew each sweet sound,
Then darted to the Sun;
Slowly the sounds came back again,
Now mixed, now one by one.

Sometimes a-dropping from the sky
I heard the sky-lark sing;
Sometimes all little birds that are,
How they seemed to fill the sea and air
With their sweet jargoning!

And now 'twas like all instruments,
Now like a lonely flute;
And now it is an angel's song,
That makes the heavens be mute.

It ceased; yet still the sails made on
A pleasant noise till noon,
A noise like of a hidden brook
In the leafy month of June,
That to the sleeping woods all night
Singeth a quiet tune.

Till noon we quietly sailed on,
Yet never a breeze did breathe:
Slowly and smoothly went the ship,
Moved onward from beneath.

Under the keel nine fathom deep,
From the land of mist and snow,
The Spirit slid: and it was he
That made the ship to go.
The sails at noon left off their tune,
And the ship stood still also.

The lonesome Spirit from the South Pole carries on the ship as far as the Line, in obedience to the angelic troop, but still requireth vengeance.

The Sun, right up above the mast,
Had fixed her to the ocean:
But in a minute she 'gan stir,

With a short uneasy motion—
Backwards and forwards half her length
With a short uneasy motion.

Then like a pawing horse let go,
She made a sudden bound:
It flung the blood into my head,
And I fell down in a swound.

The Polar Spirit's fellow-demons, the invisible inhabitants of the element, take part in his wrong; and two of them relate, one to the other, that penance long and heavy for the ancient Mariner hath been accorded to the Polar Spirit who returneth southward.

How long in that same fit I lay,
I have not to declare;
But ere my living life returned,
I heard and in my soul discerned
Two voices in the air.

"Is it he?" quoth one, "Is this the man?
By Him who died on cross,
With his cruel bow he laid full low
The harmless Albatross.

The Spirit who bideth by himself
In the land of mist and snow,
He loved the bird that loved the man
Who shot him with his bow."

The other was a softer voice,
As soft as honey-dew:
Quoth he, "The man hath penance done,
And penance more will do."

PART VI

FIRST VOICE

"But tell me, tell me! speak again,
Thy soft response renewing—
What makes that ship drive on so fast?
What is the ocean doing?"

SECOND VOICE

"Still as a slave before his lord,
The ocean hath no blast;
His great bright eye most silently
Up to the Moon is cast—

If he may know which way to go;
For she guides him smooth or grim.
See, brother, see! how graciously
She looketh down on him."

FIRST VOICE

"But why drives on that ship so fast,
Without or wave or wind?"

The Mariner hath been cast into a trance; for the angelic power causeth the vessel to drive northward faster than human life could endure.

SECOND VOICE

"The air is cut away before,
And closes from behind.

Fly, brother, fly! more high, more high!
Or we shall be belated:
For slow and slow that ship will go,
When the Mariner's trance is abated."

I woke, and we were sailing on
As in a gentle weather:
'Twas night, calm night, the Moon was high
The dead men stood together.

The supernatural motion is retarded; the Mariner awakes, and his penance begins anew.

All stood together on the deck,
For a charnel-dungeon fitter:
All fixed on me their stony eyes,
That in the Moon did glitter.

The pang, the curse, with which they died,
Had never passed away:
I could not draw my eyes from theirs,
Nor turn them up to pray.

The curse is finally expiated.

And now this spell was snapt: once more
I viewed the ocean green,
And looked far forth, yet little saw
Of what had else been seen—

Like one, that on a lonesome road
Doth walk in fear and dread,
And having once turned round, walks on,
And turns no more his head;
Because he knows, a frightful fiend
Doth close behind him tread.

But soon there breathed a wind on me,
Nor sound nor motion made:
Its path was not upon the sea.
In ripple or in shade.

It raised my hair, it fanned my cheek
Like a meadow-gale of spring—
It mingled strangely with my fears,
Yet it felt like a welcoming.

Swiftly, swiftly flew the ship,
Yet she sailed softly too:
Sweetly, sweetly blew the breeze—
On me alone it blew.

And the ancient Mariner beholdeth his native country.

Oh! dream of joy! is this indeed
The lighthouse top I see?
Is this the hill? is this the kirk?
Is this mine own countree?

We drifted o'er the harbour-bar,
And I with sobs did pray—
O let me be awake, my God!
Or let me sleep alway.

The harbour-bay was clear as glass,
So smoothly it was strewn!
And on the bay the moonlight lay,
And the shadow of the Moon.

The rock shone bright, the kirk no less,
That stands above the rock:
The moonlight steeped in silentness
The steady weathercock.

And the bay was white with silent light
Till rising from the same,
Full many shapes, that shadows were,
In crimson colours came.

The angelic spirits leave the dead bodies,

A little distance from the prow
Those crimson shadows were:
I turned my eyes upon the deck—
O Christ! what saw I there!

And appear in their own forms of light.

Each corse lay flat, lifeless and flat,
And, by the holy rood!
A man all light, a seraph-man,
On every corse there stood.

This seraph-band, each waved his hand;
It was a heavenly sight!
They stood as signals to the land,
Each one a lovely light;

This seraph-band, each waved his hand;
No voice did they impart—
No voice; but oh! the silence sank
Like music on my heart.

But soon I heard the dash of oars,
I heard the Pilot's cheer;
My head was turned perforce away
And I saw a boat appear.

The Pilot and the Pilot's boy,
I heard them coming fast:
Dear Lord in Heaven! it was a joy
The dead men could not blast.

I saw a third—I heard his voice:
It is the Hermit good!
He singeth loud his godly hymns
That he makes in the wood.
He'll shrieve my soul, he'll wash away
The Albatross's blood.

PART VII

The Hermit of the Wood.

This Hermit good lives in that wood
Which slopes down to the sea.
How loudly his sweet voice he rears!
He loves to talk with marineres
That come from a far countree.

He kneels at morn, and noon, and eve—
He hath a cushion plump:
It is the moss that wholly hides
The rotted old oak-stump.

The skiff-boat neared: I heard them talk,
"Why, this is strange, I trow!
Where are those lights so many and fair,
That signal made but now?"

Approacheth the ship with wonder.

"Strange, by my faith!" the Hermit said—
"And they answered not our cheer!
The planks look warped! and see those sails,
How thin they are and sere!
I never saw aught like to them,
Unless perchance it were

Brown skeletons of leaves that lag
My forest-brook along;
When the ivy-tod is heavy with snow,
And the owlet whoops to the wolf below,
That eats the she-wolf's young."

"Dear Lord! it hath a fiendish look—
(The Pilot made reply)
I am a-feared"—"Push on, push on!"
Said the Hermit cheerily.

The boat came closer to the ship,
But I nor spake nor stirred;
The boat came close beneath the ship,
And straight a sound was heard.

Under the water it rumbled on,
Still louder and more dread:
It reached the ship, it split the bay;
The ship went down like lead.

The ship suddenly sinketh.

Stunned by that loud and dreadful sound,
Which sky and ocean smote,
Like one that hath been seven days drowned
My body lay afloat;
But swift as dreams, myself I found
Within the Pilot's boat.

The ancient Mariner is saved in the Pilot's boat.

Upon the whirl, where sank the ship,
The boat spun round and round;
And all was still, save that the hill
Was telling of the sound.

I moved my lips—the Pilot shrieked
And fell down in a fit;
The holy Hermit raised his eyes,
And prayed where he did sit.

I took the oars: the Pilot's boy,
Who now doth crazy go,
Laughed loud and long, and all the while
His eyes went to and fro.
"Ha! ha!" quoth he, "full plain I see,
The Devil knows how to row."

And now, all in my own countree,
I stood on the firm land!
The Hermit stepped forth from the boat,
And scarcely he could stand.

The ancient Mariner earnestly entreateth the Hermit to shrieve him and the penance of life falls on him.

"O shrieve me, shrieve me, holy man!"
The Hermit crossed his brow.
"Say quick," quoth he, "I bid thee say—
What manner of man art thou?"

Forthwith this frame of mine was wrenched
With a woful agony,
Which forced me to begin my tale;
And then it left me free.

And ever and anon throughout his future life an agony constraineth him to travel from land to land,

Since then, at an uncertain hour,
That agony returns:
And till my ghastly tale is told,
This heart within me burns.

I pass, like night, from land to land;
I have strange power of speech;
That moment that his face I see,
I know the man that must hear me:
To him my tale I teach.

What loud uproar bursts from that door!
The wedding-guests are there:
But in the garden-bower the bride
And bride-maids singing are:

And hark the little vesper bell
Which biddeth me to prayer!

O Wedding-Guest! this soul hath been
Alone on a wide wide sea;
So lonely 'twas, that God himself
Scarce seemèd there to be.

O sweeter than the marriage-feast,
'Tis sweeter far to me,
To walk together to the kirk
With a goodly company!—

To walk together to the kirk,
And all together pray,
While each to his great Father bends,
Old men, and babes, and loving friends,
And youths and maidens gay!

Farewell, farewell! but this I tell
To thee, thou Wedding-Guest!
He prayeth well, who loveth well
Both man and bird and beast.

And to teach, by his own example, love and reverence to all things that God made and loveth.

He prayeth best, who loveth best
All things both great and small;
For the dear God who loveth us,
He made and loveth all.'

The Mariner, whose eye is bright,
Whose beard with age is hoar,
Is gone: and now the Wedding-Guest
Turned from the bridegroom's door.

He went like one that hath been stunned
And is of sense forlorn:
A sadder and a wiser man,
He rose the morrow morn.

SAMUEL TAYLOR COLERIDGE

The Lady of Shalott

I

On either side the river lie
Long fields of barley and of rye,
That clothe the wold and meet the sky;
And thro' the field the road runs by
    To many-tower'd Camelot;
And up and down the people go,
Gazing where the lilies blow
Round an island there below,
    The island of Shalott.

Willows whiten, aspens quiver,
Little breezes dusk and shiver
Thro' the wave that runs for ever
By the island in the river
    Flowing down to Camelot.
Four gray walls, and four gray towers,
Overlook a space of flowers,
And the silent isle imbowers
    The Lady of Shalott.

By the margin, willow-veil'd,
Slide the heavy barges trail'd
By slow horses; and unhail'd
The shallop flitteth silken-sail'd
    Skimming down to Camelot:
But who hath seen her wave her hand?
Or at the casement seen her stand?
Or is she known in all the land,
    The Lady of Shalott?

Only reapers, reaping early
In among the bearded barley,
Hear a song that echoes cheerly
From the river winding clearly,
    Down to tower'd Camelot:

And by the moon the reaper weary,
Piling sheaves in uplands airy,
Listening, whispers ''Tis the fairy
Lady of Shalott.'

II

There she weaves by night and day
A magic web with colours gay.
She has heard a whisper say,
A curse is on her if she stay
To look down to Camelot.
She knows not what the curse may be,
And so she weaveth steadily,
And little other care hath she,
The Lady of Shalott.

And moving thro' a mirror clear
That hangs before her all the year,
Shadows of the world appear.
There she sees the highway near
Winding down to Camelot:
There the river eddy whirls,
And there the surly village-churls,
And the red cloaks of market girls,
Pass onward from Shalott.

Sometimes a troop of damsels glad,
An abbot on an ambling pad,
Sometimes a curly shepherd-lad,
Or long-hair'd page in crimson clad,
Goes by to tower'd Camelot;
And sometimes thro' the mirror blue
The knights come riding two and two:
She hath no loyal knight and true,
The Lady of Shalott.

But in her web she still delights
To weave the mirror's magic sights,

For often thro' the silent nights
A funeral, with plumes and lights
        And music, went to Camelot:
Or when the moon was overhead,
Came two young lovers lately wed;
'I am half sick of shadows,' said
        The Lady of Shalott.

III

A bow-shot from her bower-eaves,
He rode between the barley-sheaves,
The sun came dazzling thro' the leaves,
And flamed upon the brazen greaves
        Of bold Sir Lancelot.
A red-cross knight for ever kneel'd
To a lady in his shield,
That sparkled on the yellow field,
        Beside remote Shalott.

The gemmy bridle glitter'd free,
Like to some branch of stars we see
Hung in the golden Galaxy.
The bridle bells rang merrily
        As he rode down to Camelot:
And from his blazon'd baldric slung
A mighty silver bugle hung,
And as he rode his armour rung,
        Beside remote Shalott.

All in the blue unclouded weather
Thick-jewell'd shone the saddle-leather,
The helmet and the helmet-feather
Burn'd like one burning flame together,
        As he rode down to Camelot.
As often thro' the purple night,
Below the starry clusters bright,
Some bearded meteor, trailing light,
        Moves over still Shalott.

His broad clear brow in sunlight glow'd;
On burnish'd hooves his war-horse trode;
From underneath his helmet flow'd
His coal-black curls as on he rode,
As he rode down to Camelot.
From the bank and from the river
He flash'd into the crystal mirror,
'Tirra lirra,' by the river
Sang Sir Lancelot.

She left the web, she left the loom,
She made three paces thro' the room,
She saw the water-lily bloom,
She saw the helmet and the plume,
She look'd down to Camelot.
Out flew the web and floated wide;
The mirror crack'd from side to side;
'The curse is come upon me!' cried
The Lady of Shalott.

IV

In the stormy east-wind straining,
The pale yellow woods were waning,
The broad stream in his banks complaining,
Heavily the low sky raining
Over tower'd Camelot;
Down she came and found a boat
Beneath a willow left afloat,
And round about the prow she wrote
The Lady of Shalott.

And down the river's dim expanse
Like some bold seër in a trance,
Seeing all his own mischance—
With a glassy countenance
Did she look to Camelot.

And at the closing of the day
She loosed the chain, and down she lay;
The broad stream bore her far away,
        The Lady of Shalott.

Lying, robed in snowy white
That loosely flew to left and right—
The leaves upon her falling light—
Thro' the noises of the night
        She floated down to Camelot:
And as the boat-head wound along
The willowy hills and fields among,
They heard her singing her last song,
        The Lady of Shalott.

Heard a carol, mournful, holy,
Chanted loudly, chanted lowly,
Till her blood was frozen slowly,
And her eyes were darken'd wholly,
        Turn'd to tower'd Camelot.
For ere she reach'd upon the tide
The first house by the water-side,
Singing in her song she died,
        The Lady of Shalott.

Under tower and balcony,
By garden-wall and gallery,
A gleaming shape she floated by,
Dead-pale between the houses high,
        Silent into Camelot.
Out upon the wharfs they came,
Knight and burgher, lord and dame,
And round the prow they read her name,
        The Lady of Shalott.

Who is this? and what is here?
And in the lighted palace near
Died the sound of royal cheer;
And they cross'd themselves for fear,
        All the knights at Camelot:

But Lancelot mused a little space;
He said, 'She has a lovely face;
God in His mercy lend her grace,
The Lady of Shalott.'

LORD TENNYSON

La Belle Dame Sans Merci

'O what can ail thee, knight-at-arms,
Alone and palely loitering?
The sedge is wither'd from the lake,
And no birds sing.

'O what can ail thee, knight-at-arms,
So haggard and so woe-begone?
The squirrel's granary is full,
And the harvest's done.

'I see a lily on thy brow
With anguish moist and fever dew,
And on thy cheek a fading rose
Fast withereth too.'

'I met a lady in the meads,
Full beautiful—a faery's child,
Her hair was long, her foot was light,
And her eyes were wild.

'I made a garland for her head,
And bracelets too, and fragrant zone;
She look'd at me as she did love,
And made sweet moan.

'I set her on my pacing steed
And nothing else saw all day long,
For sidelong would she bend, and sing
A faery's song.

'She found me roots of relish sweet
And honey wild and manna-dew,
And sure in language strange she said,
"I love thee true."

'She took me to her elfin grot,
And there she wept and sigh'd full sore,
And there I shut her wild, wild eyes
With kisses four.

'And there she lullèd me asleep,
And there I dream'd—Ah! woe betide!
The latest dream I ever dream'd
On the cold hill's side.

'I saw pale kings and princes too,
Pale warriors, death-pale were they all,
They cried—"La belle Dame sans Merci
Hath thee in thrall!"

'I saw their starved lips in the gloam
With horrid warning gapèd wide,
And I awoke and found me here
On the cold hill's side.

'And this is why I sojourn here
Alone and palely loitering,
Though the sedge is wither'd from the lake,
And no birds sing.' JOHN KEATS

The Blessèd Damozel

The blessèd damozel leaned out
From the gold bar of Heaven;
Her eyes were deeper than the depth
Of waters stilled at even;
She had three lilies in her hand,
And the stars in her hair were seven.

Her robe, ungirt from clasp to hem,
  No wrought flowers did adorn,
But a white rose of Mary's gift,
  For service meetly worn;
Her hair that lay along her back,
  Was yellow like ripe corn.

Herseemed she scarce had been a day
  One of God's choristers;
The wonder was not yet quite gone
  From that still look of hers;
Albeit, to them she left, her day
  Had counted as ten years.

(To one, it is ten years of years.
  ...Yet now, and in this place,
Surely she leaned o'er me,—her hair
  Fell all about my face...
Nothing: the autumn fall of leaves.
  The whole year sets apace.)

It was the rampart of God's house
  That she was standing on;
By God built over the sheer depth
  The which is Space begun;
So high, that looking downward thence,
  She scarce could see the sun.

It lies in Heaven, across the flood
  Of ether, as a bridge.
Beneath, the tides of day and night
  With flame and darkness ridge
The void, as low as where this earth
  Spins like a fretful midge.

Around her, lovers, newly met
  'Mid deathless love's acclaims,
Spoke evermore among themselves
  Their heart-remembered names;
And the souls mounting up to God
  Went by her like thin flames.

And still she bowed herself and stooped
  Out of the circling charm;
Until her bosom must have made
  The bar she leaned on warm,
And the lilies lay as if asleep
  Along her bended arm.

From the fixed place of Heaven she saw
  Time like a pulse shake fierce
Through all the worlds. Her gaze still strove
  Within the gulf to pierce
Its path; and now she spoke as when
  The stars sang in their spheres.

The sun was gone now; the curled moon
  Was like a little feather
Fluttering far down the gulf; and now
  She spoke through the still weather.
Her voice was like the voice the stars
  Had when they sang together.

(Ah sweet! Even now, in that bird's song,
  Strove not her accents there,
Fain to be hearkened? When those bells
  Possessed the mid-day air,
Strove not her steps to reach my side
  Down all the echoing stair?)

'I wish that he were come to me,
  For he will come,' she said.
'Have I not prayed in Heaven?—on earth,
  Lord, Lord, has he not prayed?
Are not two prayers a perfect strength?
  And shall I feel afraid?

'When round his head the aureole clings,
  And he is clothed in white,
I'll take his hand and go with him
  To the deep wells of light;
As unto a stream we will step down,
  And bathe there in God's sight.

'We two will stand beside that shrine,
Occult, withheld, untrod,
Whose lamps are stirred continually
With prayer sent up to God:
And see our old prayers, granted, melt
Each like a little cloud.

'We two will lie i' the shadow of
That living mystic tree,
Within whose secret growth the Dove
Is sometimes felt to be,
While every leaf that His plumes touch
Saith His Name audibly.

'And I myself will teach to him,
I myself, lying so,
The songs I sing here; which his voice
Shall pause in, hushed and slow,
And find some knowledge at each pause,
Or some new thing to know.'

(Alas! we two, we two, thou say'st!
Yea, one wast thou with me
That once of old. But shall God lift
To endless unity
The soul whose likeness with thy soul
Was but its love for thee?)

'We two,' she said, 'will seek the groves
Where the lady Mary is,
With her five handmaidens, whose names
Are five sweet symphonies,
Cecily, Gertrude, Magdalen,
Margaret and Rosalys.

'Circle-wise sit they, with bound locks
And foreheads garlanded;
Into the fine cloth white like flame
Weaving the golden thread,
To fashion the birth-robes for them
Who are just born, being dead.

'He shall fear, haply, and be dumb:
  Then will I lay my cheek
To his, and tell about our love,
  Not once abashed or weak:
And the dear Mother will approve
  My pride, and let me speak.

'Herself shall bring us, hand in hand,
  To Him round whom all souls
Kneel, the clear-ranged unnumbered heads
  Bowed with their aureoles:
And angels meeting us shall sing
  To their citherns and citoles.

'There will I ask of Christ the Lord
  Thus much for him and me:—
Only to live as once on earth
  With Love,—only to be,
As then awhile, for ever now
  Together, I and he.'

She gazed and listened and then said,
  Less sad of speech than mild,—
'All this is when he comes.' She ceased.
  The light thrilled towards her, filled
With angels in strong level flight.
  Her eyes prayed, and she smiled.

(I saw her smile.) But soon their path
  Was vague in distant spheres:
And then she cast her arms along
  The golden barriers,
And laid her face between her hands,
  And wept. (I heard her tears.)

DANTE GABRIEL ROSSETTI

Kubla Khan

(A VISION IN A DREAM)

In Xanadu did Kubla Khan
  A stately pleasure-dome decree:
Where Alph, the sacred river, ran
Through caverns measureless to man
  Down to a sunless sea.
So twice five miles of fertile ground
With walls and towers were girdled round:
And there were gardens bright with sinuous rills
Where blossom'd many an incense-bearing tree;
And here were forests ancient as the hills,
Enfolding sunny spots of greenery.

But oh! that deep romantic chasm which slanted
Down the green hill athwart a cedarn cover!
A savage place! as holy and enchanted
As e'er beneath a waning moon was haunted
By woman wailing for her demon-lover!
And from this chasm, with ceaseless turmoil seething
As if this earth in fast thick pants were breathing,
A mighty fountain momently was forced:
Amid whose swift half-intermitted burst
Huge fragments vaulted like rebounding hail,
Or chaffy grain beneath the thresher's flail:
And 'mid these dancing rocks at once and ever
It flung up momently the sacred river.
Five miles meandering with a mazy motion
Through wood and dale the sacred river ran,
Then reach'd the caverns measureless to man,
And sank in tumult to a lifeless ocean:
And 'mid this tumult Kubla heard from far
Ancestral voices prophesying war!

The shadow of the dome of pleasure
  Floated midway on the waves;
Where was heard the mingled measure
  From the fountain and the caves.

It was a miracle of rare device,
A sunny pleasure-dome with caves of ice!

A damsel with a dulcimer
In a vision once I saw:
It was an Abyssinian maid,
And on her dulcimer she play'd,
Singing of Mount Abora.
Could I revive within me
Her symphony and song,
To such a deep delight 'twould win me
That with music loud and long,
I would build that dome in air,
That sunny dome! Those caves of ice!
And all who heard should see them there
And all should cry, Beware! Beware!
His flashing eyes, his floating hair!
Weave a circle round him thrice,
And close your eyes with holy dread,
For he on honey-dew hath fed,
And drunk the milk of Paradise!

SAMUEL TAYLOR COLERIDGE

Eve

Eve, with her basket, was
Deep in the bells and grass,
Wading in bells and grass
Up to her knees,
Picking a dish of sweet
Berries and plums to eat,
Down in the bells and grass
Under the trees.

Mute as a mouse in a
Corner the cobra lay,
Curled round a bough of the
Cinnamon tall....

Now to get even and
Humble proud heaven and
Now was the moment or
Never at all.

'Eva!' Each syllable
Light as a flower fell,
'Eva!' he whispered the
Wondering maid,
Soft as a bubble sung
Out of a linnet's lung,
Soft and most silverly
'Eva!' he said.

Picture that orchard sprite,
Eve, with her body white,
Supple and smooth to her
Slim finger tips,
Wondering, listening,
Listening, wondering,
Eve with a berry
Half-way to her lips.

Oh had our simple Eve
Seen through the make-believe!
Had she but known the
Pretender he was!
Out of the boughs he came,
Whispering still her name,
Tumbling in twenty rings
Into the grass.

Here was the strangest pair
In the world anywhere,
Eve in the bells and grass
Kneeling, and he
Telling his story low....
Singing birds saw them go
Down the dark path to
The Blasphemous Tree.

Oh what a clatter when
Titmouse and Jenny Wren
Saw him successful and
Taking his leave!
How the birds rated him,
Now they all hated him!
How they all pitied
Poor motherless Eve!

Picture her crying
Outside in the lane,
Eve, with no dish of sweet
Berries and plums to eat,
Haunting the gate of the
Orchard in vain....
Picture the lewd delight
Under the hill to-night—
'Eva!' the toast goes round,
'Eva!' again.

RALPH HODGSON

Goblin Market

Morning and evening
Maids heard the goblins cry:
'Come buy our orchard fruits,
Come buy, come buy:
Apples and quinces,
Lemons and oranges,
Plump unpecked cherries,
Melons and raspberries,
Bloom-down-cheeked peaches,
Swart-headed mulberries,
Wild free-born cranberries,
Crab-apples, dewberries,
Pine-apples, blackberries,
Apricots, strawberries;—
All ripe together
In summer weather,—

Morns that pass by,
Fair eves that fly;
Come buy, come buy:
Our grapes fresh from the vine,
Pomegranates full and fine,
Dates and sharp bullaces,
Rare pears and greengages,
Damsons and bilberries,
Taste them and try:
Currants and gooseberries,
Bright-fire-like barberries,
Figs to fill your mouth,
Citrons from the South,
Sweet to tongue and sound to eye;
Come buy, come buy.'

Evening by evening
Among the brookside rushes,
Laura bowed her head to hear,
Lizzie veiled her blushes:
Crouching close together
In the cooling weather,
With clasping arms and cautioning lips,
With tingling cheeks and finger tips.
'Lie close,' Laura said,
Pricking up her golden head:
'We must not look at goblin men,
We must not buy their fruits:
Who knows upon what soil they fed
Their hungry thirsty roots?'
'Come buy,' call the goblins
Hobbling down the glen.
'Oh,' cried Lizzie, 'Laura, Laura,
You should not peep at goblin men.'
Lizzie covered up her eyes,
Covered close lest they should look;
Laura reared her glossy head,
And whispered like the restless brook:
'Look, Lizzie, look, Lizzie,

Down the glen tramp little men.
One hauls a basket,
One bears a plate,
One lugs a golden dish
Of many pounds weight.
How fair the vine must grow
Whose grapes are so luscious;
How warm the wind must blow
Through those fruit bushes.'
'No,' said Lizzie: 'No, no, no;
Their offers should not charm us,
Their evil gifts would harm us.'
She thrust a dimpled finger
In each ear, shut eyes and ran:
Curious Laura chose to linger
Wondering at each merchant man.
One had a cat's face,
One whisked a tail,
One tramped at a rat's pace,
One crawled like a snail,
One like a wombat prowled obtuse and furry,
One like a ratel tumbled hurry skurry.
She heard a voice like voice of doves
Cooing all together:
They sounded kind and full of loves
In the pleasant weather.

Laura stretched her gleaming neck
Like a rush-imbedded swan,
Like a lily from the beck,
Like a moonlit poplar branch,
Like a vessel at the launch
When its last restraint is gone.

Backwards up the mossy glen
Turned and trooped the goblin men
With their shrill repeated cry,
'Come buy, come buy.'
When they reached where Laura was

They stood stock still upon the moss,
Leering at each other,
Brother with queer brother;
Signalling each other,
Brother with sly brother.
One set his basket down,
One reared his plate;
One began to weave a crown
Of tendrils, leaves, and rough nuts brown
(Men sell not such in any town);
One heaved the golden weight
Of dish and fruit to offer her:
'Come buy, come buy,' was still their cry.
Laura stared but did not stir,
Longed but had no money:
The whisk-tailed merchant bade her taste
In tones as smooth as honey,
The cat-faced purr'd
The rat-paced spoke a word
Of welcome, and the snail-paced even was heard;
One parrot-voiced and jolly
Cried 'Pretty Goblin' still for 'Pretty Polly';—
One whistled like a bird.

But sweet-tooth Laura spoke in haste:
'Good folk, I have no coin;
To take were to purloin:
I have no copper in my purse,
I have no silver either,
And all my gold is on the furze
That shakes in windy weather
Above the rusty heather.'
'You have much gold upon your head.'
They answered all together:
'Buy from us with a golden curl.'
She clipped a precious golden lock,
She dropped a tear more rare than pearl,
Then sucked their fruit globes fair or red:
Sweeter than honey from the rock,

Stronger than man-rejoicing wine,
Clearer than water flowed that juice;
She never tasted such before,
How should it cloy with length of use?
She sucked and sucked and sucked the more
Fruits which that unknown orchard bore;
She sucked until her lips were sore;
Then flung the emptied rinds away
But gathered up one kernel stone,
And knew not was it night or day
As she turned home alone.

Lizzie met her at the gate
Full of wise upbraidings:
'Dear, you should not stay so late,
Twilight is not good for maidens;
Should not loiter in the glen
In the haunts of goblin men.
Do you not remember Jeanie,
How she met them in the moonlight,
Took their gifts both choice and many,
Ate their fruits and wore their flowers
Plucked from bowers
Where summer ripens at all hours?
But ever in the noonlight
She pined and pined away;
Sought them by night and day,
Found them no more but dwindled and grew grey;
Then fell with the first snow,
While to this day no grass will grow
Where she lies low:
I planted daisies there a year ago
That never blow.
You should not loiter so.'
'Nay, hush,' said Laura:
'Nay, hush, my sister:
I ate and ate my fill,
Yet my mouth waters still;
To-morrow night I will

Buy more': and kissed her:
'Have done with sorrow;
I'll bring you plums to-morrow
Fresh on their mother twigs,
Cherries worth getting;
You cannot think what figs
My teeth have met in,
What melons icy-cold
Piled on a dish of gold
Too huge for me to hold,
What peaches with a velvet nap,
Pellucid grapes without one seed:
Odorous indeed must be the mead
Whereon they grow, and pure the wave they drink
With lilies at the brink,
And sugar-sweet their sap.'

Golden head by golden head,
Like two pigeons in one nest,
Folded in each other's wings,
They lay down in their curtained bed:
Like two blossoms on one stem,
Like two flakes of new-fall'n snow,
Like two wands of ivory
Tipped with gold for awful kings.
Moon and stars gazed in at them,
Wind sang to them lullaby,
Lumbering owls forbore to fly,
Not a bat flapped to and fro
Round their nest:
Cheek to cheek and breast to breast
Locked together in one nest.

Early in the morning
When the first cock crowed his warning,
Neat like bees, as sweet and busy,
Laura rose with Lizzie:
Fetched in honey, milked the cows,
Aired and set to rights the house,

Kneaded cakes of whitest wheat,
Cakes for dainty mouths to eat,
Next churned butter, whipped up cream,
Fed their poultry, sat and sewed;
Talked as modest maidens should:
Lizzie with an open heart,
Laura in an absent dream,
One content, one sick in part;
One warbling for the mere bright day's delight,
One longing for the night.

At length slow evening came:
They went with pitchers to the reedy brook;
Lizzie most placid in her look,
Laura most like a leaping flame.
They drew the gurgling water from its deep;
Lizzie plucked purple and rich golden flags,
Then turning homewards said: 'The sunset flushes
Those furthest loftiest crags;
Come, Laura, not another maiden lags,
No wilful squirrel wags,
The beasts and birds are fast asleep.'
But Laura loitered still among the rushes
And said the bank was steep.

And said the hour was early still,
The dew not fall'n, the wind not chill:
Listening ever, but not catching
The customary cry,
'Come buy, come buy,'
With its iterated jingle
Of sugar-baited words:
Not for all her watching
Once discerning even one goblin
Racing, whisking, tumbling, hobbling;
Let alone the herds
That used to tramp along the glen,
In groups or single,
Of brisk fruit-merchant men.

Till Lizzie urged, 'O Laura, come;
I hear the fruit-call, but I dare not look:
You should not loiter longer at this brook:
Come with me home,
The stars rise, the moon bends her arc,
Each glowworm winks her spark,
Let us get home before the night grows dark:
For clouds may gather
Though this is summer weather,
Put out the lights and drench us through;
Then if we lost our way what should we do?'

Laura turned cold as stone
To find her sister heard that cry alone,
That goblin cry,
'Come buy our fruits, come buy.'
Must she then buy no more such dainty fruit?
Must she no more such succous pasture find,
Gone deaf and blind?
Her tree of life drooped from the root:
She said not one word in her heart's sore ache;
But peering thro' the dimness, nought discerning,
Trudged home, her pitcher dripping all the way;
So crept to bed, and lay
Silent till Lizzie slept;
Then sat up in a passionate yearning,
And gnashed her teeth for baulked desire, and wept
As if her heart would break.

Day after day, night after night,
Laura kept watch in vain
In sullen silence of exceeding pain.
She never caught again the goblin cry:
'Come buy, come buy'—
She never spied the goblin men
Hawking their fruits along the glen:
But when the noon waxed bright
Her hair grew thin and grey;

She dwindled, as the fair full moon doth turn
To swift decay and burn
Her fire away.

One day remembering her kernel-stone
She set it by a wall that faced the south;
Dewed it with tears, hoped for a root,
Watched for a waxing shoot,
But there came none;
It never saw the sun,
It never felt the trickling moisture run:
While with sunk eyes and faded mouth
She dreamed of melons, as a traveller sees
False waves in desert drouth
With shade of leaf-crowned trees,
And burns the thirstier in the sandful breeze.

She no more swept the house,
Tended the fowls or cows,
Fetched honey, kneaded cakes of wheat,
Brought water from the brook:
But sat down listless in the chimney-nook
And would not eat.

Tender Lizzie could not bear
To watch her sister's cankerous care
Yet not to share.
She night and morning
Caught the goblins' cry:
'Come buy our orchard fruits,
Come buy, come buy':—
Beside the brook, along the glen,
She heard the tramp of goblin men,
The voice and stir
Poor Laura could not hear;
Longed to buy fruit to comfort her,
But feared to pay too dear.

She thought of Jeanie in her grave,
Who should have been a bride;
But who for joys brides hope to have
Fell sick and died
In her gay prime,
In earliest Winter time,
With the first glazing rime,
With the first snow-fall of crisp Winter time.

Till Laura dwindling
Seemed knocking at Death's door:
Then Lizzie weighed no more
Better and worse;
But put a silver penny in her purse,
Kissed Laura, crossed the heath with clumps of furze
At twilight, halted by the brook:
And for the first time in her life
Began to listen and look.

Laughed every goblin
When they spied her peeping:
Came towards her hobbling,
Flying, running, leaping,
Puffing and blowing,
Chuckling, clapping, crowing,
Clucking and gobbling,
Mopping and mowing,
Full of airs and graces,
Pulling wry faces,
Demure grimaces,
Cat-like and rat-like,
Ratel- and wombat-like,
Snail-paced in a hurry,
Parrot-voiced and whistler,
Helter skelter, hurry skurry,
Chattering like magpies,
Fluttering like pigeons,
Gliding like fishes,—

Hugged her and kissed her:
Squeezed and caressed her:
Stretched up their dishes,
Panniers, and plates:
'Look at our apples
Russet and dun,
Bob at our cherries,
Bite at our peaches,
Citrons and dates,
Grapes for the asking,
Pears red with basking
Out in the sun,
Plums on their twigs;
Pluck them and suck them,
Pomegranates, figs.'—

'Good folk,' said Lizzie,
Mindful of Jeanie:
'Give me much and many':—
Held out her apron,
Tossed them her penny.
'Nay, take a seat with us,
Honour and eat with us,'
They answered grinning:
'Our feast is but beginning.
Night yet is early,
Warm and dew-pearly,
Wakeful and starry:
Such fruits as these
No man can carry;
Half their bloom would fly,
Half their dew would dry,
Half their flavour would pass by.
Sit down and feast with us,
Be welcome guest with us,
Cheer you and rest with us.'—
'Thank you,' said Lizzie: 'But one waits
At home alone for me:

So without further parleying,
If you will not sell me any
Of your fruits though much and many,
Give me back my silver penny
I tossed you for a fee.'—
They began to scratch their pates,
No longer wagging, purring,
But visibly demurring,
Grunting and snarling.
One called her proud,
Cross-grained, uncivil;
Their tones waxed loud,
Their looks were evil.
Lashing their tails
They trod and hustled her,
Elbowed and jostled her,
Clawed with their nails,
Barking, mewing, hissing, mocking,
Tore her gown and soiled her stocking,
Twitched her hair out by the roots,
Stamped upon her tender feet,
Held her hands and squeezed their fruits
Against her mouth to make her eat.

White and golden Lizzie stood,
Like a lily in a flood,—
Like a rock of blue-veined stone
Lashed by tides obstreperously,—
Like a beacon left alone
In a hoary roaring sea,
Sending up a golden fire,—
Like a fruit-crowned orange-tree
White with blossoms honey-sweet
Sore beset by wasp and bee,—
Like a royal virgin town
Topped with gilded dome and spire
Close beleaguered by a fleet
Mad to tug her standard down.

One may lead a horse to water,
Twenty cannot make him drink.
Though the goblins cuffed and caught her,
Coaxed and fought her,
Bullied and besought her,
Scratched her, pinched her black as ink,
Kicked and knocked her,
Mauled and mocked her,
Lizzie uttered not a word;
Would not open lip from lip
Lest they should cram a mouthful in:
But laughed in heart to feel the drip
Of juice that syrupped all her face,
And lodged in dimples of her chin,
And streaked her neck which quaked like curd.
At last the evil people
Worn out by her resistance
Flung back her penny, kicked their fruit
Along whichever road they took,
Not leaving root or stone or shoot;
Some writhed into the ground,
Some dived into the brook
With ring and ripple,
Some scudded on the gale without a sound,
Some vanished in the distance.

In a smart, ache, tingle,
Lizzie went her way;
Knew not was it night or day;
Sprang up the bank, tore thro' the furze,
Threaded copse and dingle,
And heard her penny jingle
Bouncing in her purse,—
Its bounce was music to her ear.
She ran and ran
As if she feared some goblin man
Dogged her with gibe or curse
Or something worse:

But not one goblin scurried after,
Nor was she pricked by fear;
The kind heart made her windy-paced
That urged her home quite out of breath with haste
And inward laughter.

She cried 'Laura,' up the garden,
'Did you miss me?
Come and kiss me.
Never mind my bruises,
Hug me, kiss me, suck my juices
Squeezed from goblin fruits for you,
Goblin pulp and goblin dew.
Eat me, drink me, love me;
Laura, make much of me:
For your sake I have braved the glen
And had to do with goblin merchant men.'

Laura started from her chair,
Flung her arms up in the air,
Clutched her hair:
'Lizzie, Lizzie, have you tasted
For my sake the fruit forbidden?
Must your light like mine be hidden,
Your young life like mine be wasted,
Undone in mine undoing
And ruined in my ruin,
Thirsty, cankered, goblin-ridden?'—
She clung about her sister,
Kissed and kissed and kissed her:
Tears once again
Refreshed her shrunken eyes,
Dropping like rain
After long sultry drouth;
Shaking with aguish fear, and pain,
She kissed and kissed her with a hungry mouth.

Her lips began to scorch,
That juice was wormwood to her tongue,

She loathed the feast:
Writhing as one possessed she leaped and sung,
Rent all her robe, and wrung
Her hands in lamentable haste,
And beat her breast.
Her locks streamed like the torch
Borne by a racer at full speed,
Or like the mane of horses in their flight,
Or like an eagle when she stems the light
Straight toward the sun,
Or like a caged thing freed,
Or like a flying flag when armies run.

Swift fire spread through her veins, knocked at her heart,
Met the fire smouldering there
And overbore its lesser flame;
She gorged on bitterness without a name:
Ah! fool, to choose such part
Of soul-consuming care!
Sense failed in the mortal strife:
Like the watch-tower of a town
Which an earthquake shatters down,
Like a lightning-stricken mast,
Like a wind-uprooted tree
Spun about,
Like a foam-topped waterspout
Cast down headlong in the sea,
She fell at last;
Pleasure past and anguish past,
Is it death or is it life?

Life out of death.
That night long Lizzie watched by her,
Counted her pulse's flagging stir,
Felt for her breath,
Held water to her lips, and cooled her face
With tears and fanning leaves:
But when the first birds chirped about their eaves,

And early reapers plodded to the place
Of golden sheaves,
And dew-wet grass
Bowed in the morning winds so brisk to pass,
And new buds with new day
Opened of cup-like lilies on the stream,
Laura awoke as from a dream,
Laughed in the innocent old way,
Hugged Lizzie but not twice or thrice;
Her gleaming locks showed not one thread of grey,
Her breath was sweet as May
And light danced in her eyes.

Days, weeks, months, years
Afterwards, when both were wives
With children of their own;
Their mother-hearts beset with fears,
Their lives bound up in tender lives;
Laura would call the little ones
And tell them of her early prime,
Those pleasant days long gone
Of not-returning time:
Would talk about the haunted glen,
The wicked, quaint fruit-merchant men,
Their fruits like honey to the throat
But poison in the blood;
(Men sell not such in any town:)
Would tell them how her sister stood
In deadly peril to do her good,
And win the fiery antidote:
Then joining hands to little hands
Would bid them cling together,
'For there is no friend like a sister
In calm or stormy weather;
To cheer one on the tedious way,
To fetch one if one goes astray,
To lift one if one totters down,
To strengthen whilst one stands.'

CHRISTINA ROSSETTI

The Forsaken Merman

Come, dear children, let us away;
Down and away below,
Now my brothers call from the bay;
Now the great winds shoreward blow;
Now the salt tides seaward flow;
Now the wild white horses play,
Champ and chafe and toss in the spray.
Children dear, let us away.
This way, this way.

Call her once before you go.
Call once yet.
In a voice that she will know:
'Margaret! Margaret!'
Children's voices should be dear
(Call once more) to a mother's ear:
Children's voices, wild with pain.
Surely she will come again.

Call her once and come away.
This way, this way.
'Mother dear, we cannot stay.'
The wild white horses foam and fret.
Margaret! Margaret!

Come, dear children, come away down.
Call no more.
One last look at the white-walled town,
And the little grey church on the windy shore.
Then come down.
She will not come though you call all day.
Come away, come away.

Children dear, was it yesterday
We heard the sweet bells over the bay?
In the caverns where we lay,
Through the surf and through the swell,
The far-off sound of a silver bell?

Sand-strewn caverns, cool and deep,
Where the winds are all asleep;
Where the spent lights quiver and gleam;
Where the salt weed sways in the stream;
Where the sea-beasts ranged all round
Feed in the ooze of their pasture-ground;
Where the sea-snakes coil and twine,
Dry their mail and bask in the brine;
Where great whales come sailing by,
Sail and sail, with unshut eye,
Round the world for ever and aye?
When did music come this way?
Children dear, was it yesterday?

Children dear, was it yesterday
(Call yet once) that she went away?
Once she sate with you and me,
On a red gold throne in the heart of the sea,
And the youngest sate on her knee.
She combed its bright hair, and she tended it well,
When down swung the sound of the far-off bell.
She sighed, she looked up through the clear green sea.
She said; 'I must go, for my kinsfolk pray
In the little grey church on the shore to-day.
'Twill be Easter-time in the world—ah me!
And I lose my poor soul, Merman, here with thee.'
I said; 'Go up, dear heart, through the waves;
Say thy prayer, and come back to the kind sea-caves.'
She smiled, she went up through the surf in the bay.
Children dear, was it yesterday?

Children dear, were we long alone?
'The sea grows stormy, the little ones moan.
Long prayers,' I said, 'in the world they say.
Come,' I said, and we rose through the surf in the bay.
We went up the beach, by the sandy down
Where the sea-stocks bloom, to the white-walled town.
Through the narrow paved streets, where all was still,

To the little grey church on the windy hill.
From the church came a murmur of folk at their prayers,
But we stood without in the cold blowing airs,
We climbed on the graves, on the stones, worn with rains,
And we gazed up the aisle through the small leaded panes.
She sate by the pillar; we saw her clear:
'Margaret, hist! come quick, we are here.
Dear heart,' I said, 'we are long alone.
The sea grows stormy, the little ones moan.'
But, ah, she gave me never a look,
For her eyes were sealed to the holy book.
Loud prays the priest; shut stands the door.
Come away, children, call no more.
Come away, come down, call no more.

Down, down, down.
Down to the depths of the sea.
She sits at her wheel in the humming town,
Singing most joyfully.
Hark, what she sings: 'O joy, O joy,
For the humming street, and the child with its toy.
For the priest, and the bell, and the holy well.
For the wheel where I spun,
And the blessed light of the sun.'
And so she sings her fill,
Singing most joyfully,
Till the shuttle falls from her hand,
And the whizzing wheel stands still.
She steals to the window, and looks at the sand;
And over the sand at the sea;
And her eyes are set in a stare;
And anon there breaks a sigh,
And anon there drops a tear,
From a sorrow-clouded eye,
And a heart sorrow-laden,
A long, long sigh,
For the cold strange eyes of a little Mermaiden,
And the gleam of her golden hair.

Come away, away children.
Come children, come down.
The hoarse wind blows colder;
Lights shine in the town.
She will start from her slumber
When gusts shake the door;
She will hear the winds howling,
Will hear the waves roar.
We shall see, while above us
The waves roar and whirl,
A ceiling of amber,
A pavement of pearl.
Singing, 'Here came a mortal,
But faithless was she.
And alone dwell for ever
The kings of the sea.'

But, children, at midnight,
When soft the winds blow;
When clear falls the moonlight;
When spring-tides are low:
When sweet airs come seaward
From heaths starred with broom;
And high rocks throw mildly
On the blanched sands a gloom:
Up the still, glistening beaches,
Up the creeks we will hie;
Over banks of bright seaweed
The ebb-tide leaves dry.
We will gaze, from the sand-hills,
At the white, sleeping town;
At the church on the hill-side—
And then come back down.
Singing, 'There dwells a loved one,
But cruel is she.
She left lonely for ever
The kings of the sea.'

MATTHEW ARNOLD

The Listeners

'Is there anybody there?' said the Traveller,
  Knocking on the moonlit door;
And his horse in the silence champ'd the grasses
  Of the forest's ferny floor:
And a bird flew up out of the turret,
  Above the Traveller's head:
And he smote upon the door again a second time;
  'Is there anybody there?' he said.
But no one descended to the Traveller;
  No head from the leaf-fringed sill
Lean'd over and look'd into his grey eyes,
  Where he stood perplex'd and still.
But only a host of phantom listeners
  That dwelt in the lone house then
Stood listening in the quiet of the moonlight
  To that voice from the world of men:
Stood thronging the faint moonbeams on the dark stair,
  That goes down to the empty hall,
Hearkening in an air stirr'd and shaken
  By the lonely Traveller's call.
And he felt in his heart their strangeness,
  Their stillness answering his cry,
While his horse moved, cropping the dark turf,
  'Neath the starr'd and leafy sky;
For he suddenly smote on the door, even
  Louder, and lifted his head:—
'Tell them I came, and no one answer'd,
  That I kept my word,' he said.
Never the least stir made the listeners,
  Though every word he spake
Fell echoing through the shadowiness of the still house
  From the one man left awake:
Ay, they heard his foot upon the stirrup,
  And the sound of iron on stone,
And how the silence surged softly backward,
  When the plunging hoofs were gone.

WALTER DE LA MARE

Flannan Isle

'Though three men dwell on Flannan Isle
To keep the lamp alight,
As we steer'd under the lee, we caught
No glimmer through the night.'

A passing ship at dawn had brought
The news; and quickly we set sail,
To find out what strange thing might ail
The keepers of the deep-sea light.

The winter day broke blue and bright,
With glancing sun and glancing spray,
As o'er the swell our boat made way,
As gallant as a gull in flight.

But, as we near'd the lonely Isle;
And look'd up at the naked height;
And saw the lighthouse towering white,
With blinded lantern, that all night
Had never shot a spark
Of comfort through the dark,
So ghostly in the cold sunlight
It seem'd, that we were struck the while
With wonder all too dread for words.

And, as into the tiny creek
We stole beneath the hanging crag,
We saw three queer, black, ugly birds—
Too big, by far, in my belief,
For guillemot or shag—
Like seamen sitting bolt-upright
Upon a half-tide reef:
But, as we near'd, they plunged from sight,
Without a sound, or spurt of white.

And still too 'mazed to speak,
We landed; and made fast the boat;

shag] crested cormorant.

And climb'd the track in single file,
Each wishing he was safe afloat,
On any sea, however far,
So it be far from Flannan Isle:
And still we seem'd to climb, and climb,
As though we'd lost all count of time,
And so must climb for evermore.
Yet, all too soon, we reached the door—
The black, sun-blister'd lighthouse-door,
That gaped for us ajar.

As, on the threshold, for a spell,
We paused, we seem'd to breathe the smell
Of limewash and of tar,
Familiar as our daily breath,
As though 'twere some strange scent of death:
And so, yet wondering, side by side,
We stood a moment, still tongue-tied:
And each with black foreboding eyed
The door, ere we should fling it wide,
To leave the sunlight for the gloom:
Till, plucking courage up, at last,
Hard on each other's heels we pass'd
Into the living-room.

Yet, as we crowded through the door,
We only saw a table, spread
For dinner, meat and cheese and bread;
But all untouch'd; and no one there:
As though, when they sat down to eat,
Ere they could even taste,
Alarm had come; and they in haste
Had risen and left the bread and meat:
For at the table-head a chair
Lay tumbled on the floor.

We listen'd; but we only heard
The feeble cheeping of a bird
That starved upon its perch:

And, listening still, without a word,
We set about our hopeless search.
We hunted high, we hunted low;
And soon ransack'd the empty house;
Then o'er the Island, to and fro,
We ranged, to listen and to look
In every cranny, cleft or nook
That might have hid a bird or mouse:
But, though we search'd from shore to shore,
We found no sign in any place:
And soon again stood face to face
Before the gaping door:
And stole into the room once more
As frighten'd children steal.

Aye: though we hunted high and low,
And hunted everywhere,
Of the three men's fate we found no trace
Of any kind in any place,
But a door ajar, and an untouch'd meal,
And an overtoppled chair.
And, as we listen'd in the gloom
Of that forsaken living-room—
A chill clutch on our breath—
We thought how ill-chance came to all
Who kept the Flannan Light:
And how the rock had been the death
Of many a likely lad:
How six had come to a sudden end,
And three had gone stark mad:
And one whom we'd all known as friend
Had leapt from the lantern one still night,
And fallen dead by the lighthouse wall:
And long we thought
On the three we sought,
And of what might yet befall.

Like curs, a glance has brought to heel,
We listen'd, flinching there:

And look'd, and look'd, on the untouched meal,
And the overtoppled chair.
We seem'd to stand for an endless while,
Though still no word was said,
Three men alive on Flannan Isle,
Who thought on three men dead.

W. W. GIBSON

Welsh Incident

'But that was nothing to what things came out
From the sea-caves of Criccieth yonder.'
'What were they? Mermaids? dragons? ghosts?'
'Nothing at all of any things like that.'
'What were they, then?'
'All sorts of queer things,
Things never seen or heard or written about,
Very strange, un-Welsh, utterly peculiar
Things. Oh, solid enough they seemed to touch,
Had anyone dared it. Marvellous creation,
All various shapes and sizes and no sizes,
All new, each perfectly unlike his neighbour,
Though all came moving slowly out together.'
'Describe just one of them.'
'I am unable.'
'What were their colours?'
'Mostly nameless colours,
Colours you'd like to see; but one was puce
Or perhaps more like crimson, but not purplish.
Some had no colour.'
'Tell me, had they legs?'
'Not a leg or foot among them that I saw.'
'But did these things come out in any order?
What o'clock was it? What was the day of the week?
Who else was present? What was the weather?'
'I was coming to that. It was half-past three
On Easter Tuesday last. The sun was shining.
The Harlech Silver Band played *Marchog Jesu*
On thirty-seven shimmering instruments,

Collecting for Carnarvon's (Fever) Hospital Fund.
The populations of Pwllheli, Criccieth,
Portmadoc, Borth, Tremadoc, Penrhyndeudraeth,
Were all assembled. Criccieth's mayor addressed them
First in good Welsh and then in fluent English,
Twisting his fingers in his chain of office,
Welcoming the things. They came out on the sand,
Not keeping time to the band, moving seaward
Silently at a snail's pace. But at last
The most odd, indescribable thing of all
Which hardly one man there could see for wonder
Did something recognizably a something.'
'Well, what?'
'It made a noise.'
'A frightening noise?'
'No, no.'
'A musical noise? A noise of scuffling?'
'No, but a very loud, respectable noise—
Like groaning to oneself on Sunday morning
In Chapel, close before the second psalm.'
'What did the mayor do?'
'I was coming to that.'

ROBERT GRAVES

The Way through the Woods

They shut the road through the woods
Seventy years ago.
Weather and rain have undone it again,
And now you would never know
There was once a road through the woods
Before they planted the trees.
It is underneath the coppice and heath,
And the thin anemones.
Only the keeper sees
That, where the ring-dove broods,
And the badgers roll at ease,
There was once a road through the woods.

Yet, if you enter the woods
  Of a summer evening late,
When the night-air cools on the trout-ringed pools
  Where the otter whistles his mate.
(They fear not men in the woods
  Because they see so few)
You will hear the beat of a horse's feet
  And the swish of a skirt in the dew,
  Steadily cantering through
The misty solitudes,
  As though they perfectly knew
The old lost road through the woods...
But there is no road through the woods!

RUDYARD KIPLING

Outlaws

Owls: they whinney down the night,
  Bats go zigzag by.
Ambushed in shadow out of sight
  The outlaws lie.

Old gods, shrunk to shadows, there
  In the wet woods they lurk,
Greedy of human stuff to snare
  In webs of murk.

Look up, else your eye must drown
  In a moving sea of black
Between the tree-tops, upside down
  Goes the sky-track.

Look up, else your feet will stray
  Towards that dim ambuscade,
Where spider-like they catch their prey
  In nets of shade.

For though creeds whirl away in dust,
  Faith fails and men forget,
These aged gods of fright and lust
  Cling to life yet.

Old gods almost dead, malign,
  Starved of their ancient dues,
Incense and fruit, fire, blood and wine
  And an unclean muse.

Banished to woods and a sickly moon,
  Shrunk to mere bogey things,
Who spoke with thunder once at noon
  To prostrate kings.

With thunder from an open sky
  To peasant, tyrant, priest,
Bowing in fear with a dazzled eye
  Towards the East.

Proud gods, humbled, sunk so low,
  Living with ghosts and ghouls,
And ghosts of ghosts and last year's snow
  And dead toadstools.

ROBERT GRAVES

SECTION 4

THE ETERNAL THEME

When man reaches certain levels of civilization his feelings may become still further complicated by association, not only with instinctive impulses and with cherished ideas but with values which may be completely impersonal. We then see the emergence of the highest ideals or, as we sometimes call them, the supreme values, such as beauty, truth, and goodness.

These values were praised by the Greeks, and form a part of all great religions. The search for beauty, truth, and goodness has been for mankind the quest of the Holy Grail. Poets in their worship of beauty in nature, art, and the lives of men have glimpses of that truth which makes the universe a unity; and some have found that truth and beauty are the same:

> Beauty is truth, truth beauty,—that is all
> Ye know on earth, and all ye need to know.

These ideals have created new and better ways of life, and the feeling for goodness has been a constant inspiration:

> Only a sweet and virtuous soul,
> Like season'd timber, never gives;

and the poet praises the man who

> makes the heaven his book,
> His wisdom heavenly things.

When a poet feels intensely about life and this unintelligible world his thoughts turn also to death, which is an even greater mystery, perhaps a greater adventure than life.

Now we should expect a poem on death to be sad, and because it is sad we may ask how we can be said to enjoy it.

Poems on death may be sad but they are not therefore depressing. The poet must turn his feelings of sorrow at the death of a friend into something more than a mere statement of his loss.

In writing his poem he is creating a work of art, controlling and ordering his words to become a pattern of imagery and sound; and this work of art has the power to excite, and to expand in the mind. The words may be so charged with meaning that we may not begin fully to understand a great poem for ten years or twenty. But we can grow towards it, finding ever greater felicities of phrase and significance which give that overplus of pleasure we find in all art. Such perfection of expression is found in the poem beginning:

> Ah, Sun-flower! weary of time,
> Who countest the steps of the sun;
> Seeking after that sweet golden clime,
> Where the traveller's journey is done.

So we may read about death without dejection and enjoy such a poem as this, in which the poet's sorrow is transformed into an expression of serene acceptance:

> Fear no more the heat o' the sun,
>   Nor the furious winter's rages;
> Thou thy worldly task hast done,
>   Home art gone and ta'en thy wages:
> Golden lads and girls all must,
> As chimney-sweepers, come to dust.

The Perfect Life

It is not growing like a tree
In bulk, doth make Man better be;
Or standing long an oak, three hundred year,
To fall a log at last, dry, bald, and sere:
A lily of a day
Is fairer far in May,
Although it fall and die that night—
It was the plant and flower of Light.
In small proportions we just beauties see;
And in short measures life may perfect be.

BEN JONSON

Sweet Content

Art thou poor, yet hast thou golden slumbers?
O sweet content!
Art thou rich, yet is thy mind perplex'd?
O punishment!
Dost thou laugh to see how fools are vex'd
To add to golden numbers golden numbers?
O sweet content! O sweet, O sweet content!
Work apace, apace, apace, apace;
Honest labour bears a lovely face;
Then hey nonny nonny—hey nonny nonny!

Canst drink the waters of the crispèd spring?
O sweet content!
Swim'st thou in wealth, yet sink'st in thine own tears?
O punishment!
Then he that patiently want's burden bears,
No burden bears, but is a king, a king!
O sweet content! O sweet, O sweet content!
Work apace, apace, apace, apace;
Honest labour bears a lovely face;
Then hey nonny nonny—hey nonny nonny!

THOMAS DEKKER

The Character of a Happy Life

How happy is he born and taught
That serveth not another's will;
Whose armour is his honest thought,
And simple truth his utmost skill!

Whose passions not his masters are;
Whose soul is still prepared for death,
Untied unto the world by care
Of public fame or private breath;

Who envies none that chance doth raise,
Nor vice; who never understood
How deepest wounds are given by praise;
Nor rules of state, but rules of good;

Who hath his life from rumours freed;
Whose conscience is his strong retreat;
Whose state can neither flatterers feed,
Nor ruin make oppressors great;

Who God doth late and early pray
More of His grace than gifts to lend;
And entertains the harmless day
With a religious book or friend;

—This man is freed from servile bands
Of hope to rise or fear to fall:
Lord of himself, though not of lands,
And having nothing, yet hath all.

SIR HENRY WOTTON

Integer Vitae

The man of life upright,
　Whose guiltless heart is free
From all dishonest deeds,
　Or thought of vanity;

The man whose silent days
  In harmless joys are spent,
Whom hopes cannot delude,
  Nor sorrow discontent;

That man needs neither towers
  Nor armour for defence,
Nor secret vaults to fly
  From thunder's violence:

He only can behold
  With unaffrighted eyes
The horrors of the deep
  And terrors of the skies.

Thus, scorning all the cares
  That fate or fortune brings,
He makes the heaven his book,
  His wisdom heavenly things;

Good thoughts his only friends,
  His wealth a well-spent age,
The earth his sober inn
  And quiet pilgrimage.

THOMAS CAMPION

To Daffodils

Fair Daffodils, we weep to see
    You haste away so soon:
As yet the early-rising Sun
    Has not attain'd his noon.
        Stay, stay,
    Until the hasting day
        Has run
    But to the even-song;
And, having pray'd together, we
    Will go with you along.

We have short time to stay, as you,
We have as short a Spring!
As quick a growth to meet decay
As you, or any thing.
We die,
As your hours do, and dry
Away
Like to the Summer's rain;
Or as the pearls of morning's dew
Ne'er to be found again.

ROBERT HERRICK

The Poplar Field

The poplars are fell'd! farewell to the shade
And the whispering sound of the cool colonnade;
The winds play no longer and sing in the leaves,
Nor Ouse on his bosom their image receives.

Twelve years have elapsed since I last took a view
Of my favourite field, and the bank where they grew;
And now in the grass behold they are laid,
And the tree is my seat that once lent me a shade!

The blackbird has fled to another retreat
Where the hazels afford him a screen from the heat,
And the scene where his melody charm'd me before
Resounds with his sweet-flowing ditty no more.

My fugitive years are all hasting away,
And I must ere long lie as lowly as they,
With a turf on my breast and a stone at my head,
Ere another such grove shall arise in its stead.

The change both my heart and my fancy employs,
I reflect on the frailty of man and his joys;
Short-lived as we are, yet our pleasures, we see,
Have a still shorter date, and die sooner than we.

WILLIAM COWPER

The Flower-fed Buffaloes

The flower-fed buffaloes of the spring
In the days of long ago,
Ranged where the locomotives sing
And the prairie flowers lie low:—
The tossing, blooming, perfumed grass
Is swept away by the wheat,
Wheels and wheels and wheels spin by
In the spring that still is sweet.
But the flower-fed buffaloes of the spring
Left us, long ago.
They gore no more, they bellow no more,
They trundle around the hills no more:—
With the Blackfeet, lying low,
With the Pawnees, lying low,
Lying low. VACHEL LINDSAY

Death the Leveller

The glories of our blood and state
  Are shadows, not substantial things;
There is no armour against Fate;
  Death lays his icy hand on kings:
    Sceptre and crown
    Must tumble down,
And in the dust be equal made
With the poor crookèd scythe and spade.

Some men with swords may reap the field,
  And plant fresh laurels where they kill;
But their strong nerves at last must yield;
  They tame but one another still:
    Early or late
    They stoop to fate,
And must give up their murmuring breath
When they, pale captives, creep to death.

The garlands wither on your brow;
  Then boast no more your mighty deeds!
Upon Death's purple altar now
  See where the victor-victim bleeds.
        Your heads must come
        To the cold tomb:
Only the actions of the just
Smell sweet and blossom in their dust.

JAMES SHIRLEY

Virtue

Sweet day, so cool, so calm, so bright,
The bridal of the earth and sky:
The dew shall weep thy fall to-night;
        For thou must die.

Sweet rose, whose hue, angry and brave
Bids the rash gazer wipe his eye,
Thy root is ever in its grave,
        And thou must die.

Sweet spring, full of sweet days and roses,
A box where sweets compacted lie,
My music shows ye have your closes,
        And all must die.

Only a sweet and virtuous soul,
Like season'd timber, never gives;
But though the whole world turn to coal,
        Then chiefly lives.

GEORGE HERBERT

On the Tombs in Westminster Abbey

Mortality, behold and fear!
What a change of flesh is here!
Think how many royal bones
Sleep within this heap of stones:

Here they lie had realms and lands,
Who now want strength to stir their hands:
Where from their pulpits seal'd with dust
They preach, 'In greatness is no trust.'
Here's an acre sown indeed
With the richest, royall'st seed
That the earth did e'er suck in
Since the first man died for sin:
Here the bones of birth have cried—
'Though gods they were, as men they died.'
Here are sands, ignoble things,
Dropt from the ruin'd sides of kings;
Here's a world of pomp and state,
Buried in dust, once dead by fate.

FRANCIS BEAUMONT

Elegy Written in a Country Churchyard

The Curfew tolls the knell of parting day,
The lowing herd wind slowly o'er the lea,
The plowman homeward plods his weary way,
And leaves the world to darkness and to me.

Now fades the glimmering landscape on the sight,
And all the air a solemn stillness holds,
Save where the beetle wheels his droning flight,
And drowsy tinklings lull the distant folds;

Save that from yonder ivy-mantled tow'r
The moping owl does to the moon complain
Of such, as wand'ring near her secret bow'r,
Molest her ancient solitary reign.

Beneath those rugged elms, that yew-tree's shade,
Where heaves the turf in many a mould'ring heap,
Each in his narrow cell for ever laid,
The rude Forefathers of the hamlet sleep.

The breezy call of incense-breathing Morn,
  The swallow twitt'ring from the straw-built shed,
The cock's shrill clarion, or the echoing horn,
  No more shall rouse them from their lowly bed.

For them no more the blazing hearth shall burn,
  Or busy housewife ply her evening care:
No children run to lisp their sire's return,
  Or climb his knees the envied kiss to share.

Oft did the harvest to their sickle yield,
  Their furrow oft the stubborn glebe has broke;
How jocund did they drive their team afield!
  How bow'd the woods beneath their sturdy stroke!

Let not Ambition mock their useful toil,
  Their homely joys, and destiny obscure;
Nor Grandeur hear with a disdainful smile
  The short and simple annals of the poor.

The boast of heraldry, the pomp of pow'r,
  And all that beauty, all that wealth e'er gave,
Awaits alike th' inevitable hour.
  The paths of glory lead but to the grave.

Nor you, ye Proud, impute to These the fault,
  If Mem'ry o'er their Tomb no Trophies raise,
Where thro' the long-drawn aisle and fretted vault
  The pealing anthem swells the note of praise.

Can storied urn or animated bust
  Back to its mansion call the fleeting breath?
Can Honour's voice provoke the silent dust,
  Or Flatt'ry soothe the dull cold ear of Death?

Perhaps in this neglected spot is laid
  Some heart once pregnant with celestial fire;
Hands, that the rod of empire might have sway'd,
  Or wak'd to ecstasy the living lyre.

But Knowledge to their eyes her ample page
  Rich with the spoils of time did ne'er unroll;
Chill Penury repress'd their noble rage,
  And froze the genial current of the soul.

Full many a gem of purest ray serene,
  The dark unfathom'd caves of ocean bear:
Full many a flower is born to blush unseen
  And waste its sweetness on the desert air.

Some village-Hampden that with dauntless breast
  The little tyrant of his fields withstood;
Some mute inglorious Milton here may rest,
  Some Cromwell guiltless of his country's blood.

Th' applause of list'ning senates to command,
  The threats of pain and ruin to despise,
To scatter plenty o'er a smiling land,
  And read their history in a nation's eyes.

Their lot forbade: nor circumscribed alone
  Their growing virtues, but their crimes confined;
Forbade to wade through slaughter to a throne,
  And shut the gates of mercy on mankind.

The struggling pangs of conscious truth to hide,
  To quench the blushes of ingenuous shame,
Or heap the shrine of Luxury and Pride
  With incense kindled at the Muse's flame.

Far from the madding crowd's ignoble strife,
  Their sober wishes never learn'd to stray;
Along the cool sequester'd vale of life
  They kept the noiseless tenor of their way.

Yet ev'n these bones from insult to protect
  Some frail memorial still erected nigh,
With uncouth rhymes and shapeless sculpture deck'd,
  Implores the passing tribute of a sigh.

Their name, their years, spelt by th' unletter'd muse,
  The place of fame and elegy supply:
And many a holy text around she strews,
  That teach the rustic moralist to die.

For who, to dumb Forgetfulness a prey,
  This pleasing anxious being e'er resign'd,
Left the warm precincts of the cheerful day,
  Nor cast one longing ling'ring look behind?

On some fond breast the parting soul relies,
  Some pious drops the closing eye requires;
Ev'n from the tomb the voice of Nature cries,
  Ev'n in our Ashes live their wonted Fires.

For thee, who mindful of th' unhonour'd dead
  Dost in these lines their artless tale relate;
If chance, by lonely contemplation led,
  Some kindred spirit shall inquire thy fate,

Haply some hoary-headed Swain may say,
  'Oft have we seen him at the peep of dawn
Brushing with hasty steps the dews away
  To meet the sun upon the upland lawn.

'There at the foot of yonder nodding beech
  That wreathes its old fantastic roots so high,
His listless length at noontide would he stretch,
  And pore upon the brook that babbles by.

'Hard by yon wood, now smiling as in scorn,
  Mutt'ring his wayward fancies he would rove,
Now drooping, woeful wan, like one forlorn,
  Or craz'd with care, or cross'd in hopeless love.

'One morn I miss'd him on the custom'd hill,
  Along the heath and near his fav'rite tree;
Another came; not yet beside the rill,
  Nor up the lawn, nor at the wood was he;

'The next with dirges due in sad array
  Slow thro' the church-way path we saw him borne.
Approach and read (for thou canst read) the lay,
  Grav'd on the stone beneath yon aged thorn.'

THE EPITAPH

Here rests his head upon the lap of Earth
  A youth to Fortune and to Fame unknown.
Fair Science frown'd not on his humble birth,
  And Melancholy mark'd him for her own.

Large was his bounty, and his soul sincere,
  Heav'n did a recompense as largely send:
He gave to Mis'ry all he had, a tear;
  He gain'd from Heaven ('twas all he wish'd) a friend.

No farther seek his merits to disclose,
  Or draw his frailties from their dread abode,
(There they alike in trembling hope repose,)
  The bosom of his Father and his God.

THOMAS GRAY

A Widow Bird

A widow bird sate mourning for her love
  Upon a wintry bough;
The frozen wind crept on above,
  The freezing stream below.

There was no leaf upon the forest bare.
  No flower upon the ground,
And little motion in the air
  Except the mill-wheel's sound.

P. B. SHELLEY

Patch-Shaneen

Shaneen and Maurya Prendergast
Lived west in Carnareagh,
And they'd a cur-dog, cabbage plot,
A goat, and cock of hay.

He was five foot one or two,
Herself was four foot ten,
And he went travelling asking meal
Above through Caragh Glen.

She'd pick her bag of carrageen
Or perries through the surf,
Or loan an ass of Foxy Jim
To fetch her creel of turf.

Till on one windy Samhain night,
When there's stir among the dead,
He found her perished, stiff and stark,
Beside him in the bed.

And now when Shaneen travels far
From Droum to Ballyhyre
The women lay him sacks of straw,
Beside the seed of fire.

And when the grey cocks crow and flap
And winds are in the sky,
'Oh, Maurya, Maurya, are you dead?'
You'll hear Patch-Shaneen cry.

J. M. SYNGE

Proud Maisie

Proud Maisie is in the wood,
  Walking so early;
Sweet Robin sits on the bush,
  Singing so rarely.

'Tell me, thou bonny bird,
  When shall I marry me?'
'When six braw gentlemen
  Kirkward shall carry ye.'

'Who makes the bridal bed,
  Birdie, say truly?'
'The grey-headed sexton
  That delves the grave duly.

'The glow-worm o'er grave and stone
  Shall light thee steady.
The owl from the steeple sing,
  "Welcome, proud lady."'

SIR WALTER SCOTT

The Song of the Mad Prince

Who said, 'Peacock Pie'?
  The old King to the sparrow:
Who said, 'Crops are ripe'?
  Rust to the harrow:
Who said, 'Where sleeps she now?
  Where rests she now her head,
Bathed in Eve's loveliness'?
  That's what I said.

Who said, 'Ay, mum's the word'?
  Sexton to willow:
Who said, 'Green dust for dreams,
  Moss for a pillow'?
Who said, 'All Time's delight
  Hath she for narrow bed;
Life's troubled bubble broken'?—
  That's what I said.

WALTER DE LA MARE

Break, Break, Break

Break, break, break,
  On thy cold grey stones, O Sea!
And I would that my tongue could utter
  The thoughts that arise in me.

O well for the fisherman's boy,
  That he shouts with his sister at play!
O well for the sailor lad,
  That he sings in his boat on the bay!

And the stately ships go on
  To their haven under the hill;
But O for the touch of a vanished hand,
  And the sound of a voice that is still!

Break, break, break,
  At the foot of thy crags, O Sea!
But the tender grace of a day that is dead
  Will never come back to me.

LORD TENNYSON

A Slumber did my Spirit Seal

A slumber did my spirit seal;
  I had no human fears:
She seemed a thing that could not feel
  The touch of earthly years.

No motion has she now, no force;
  She neither hears nor sees,
Rolled round in earth's diurnal course,
  With rocks, and stones, and trees.

WILLIAM WORDSWORTH

Requiescat

Strew on her roses, roses,
  And never a spray of yew.
In quiet she reposes:
  Ah! would that I did too.

Her mirth the world required;
  She bathed it in smiles of glee.
But her heart was tired, tired,
  And now they let her be.

Her life was turning, turning,
  In mazes of heat and sound.
But for peace her soul was yearning,
  And now peace laps her round.

Her cabin'd, ample Spirit,
  It flutter'd and fail'd for breath.
To-night it doth inherit
  The vasty hall of Death.

MATTHEW ARNOLD

Prospice

Fear death?—to feel the fog in my throat,
  The mist in my face,
When the snows begin, and the blasts denote
  I am nearing the place,
The power of the night, the press of the storm,
  The post of the foe;
Where he stands, the Arch Fear in a visible form,
  Yet the strong man must go:
For the journey is done and the summit attained,
  And the barriers fall,
Though a battle's to fight ere the guerdon be gained,
  The reward of it all.
I was ever a fighter, so—one fight more,
  The best and the last!
I would hate that death bandaged my eyes, and forbore,
  And bade me creep past.
No! let me taste the whole of it, fare like my peers
  The heroes of old,
Bear the brunt, in a minute pay glad life's arrears
  Of pain, darkness and cold.

For sudden the worst turns the best to the brave,
    The black minute's at end,
And the elements' rage, the fiend-voices that rave,
    Shall dwindle, shall blend,
Shall change, shall become first a peace out of pain,
    Then a light, then thy breast,
O thou soul of my soul! I shall clasp thee again,
    And with God be the rest!

ROBERT BROWNING

Up-Hill

Does the road wind up-hill all the way?
  Yes, to the very end.
Will the day's journey take the whole long day?
  From morn to night, my friend.

But is there for the night a resting-place?
  A roof for when the slow, dark hours begin.
May not the darkness hide it from my face?
  You cannot miss that inn.

Shall I meet other wayfarers at night?
  Those who have gone before.
Then must I knock, or call when just in sight?
  They will not keep you standing at that door.

Shall I find comfort, travel-sore and weak?
  Of labour you shall find the sum.
Will there be beds for me and all who seek?
  Yea, beds for all who come.

CHRISTINA ROSSETTI

Prognosis

Good-bye, Winter
The days are getting longer,
The tea-leaf in the teacup
Is herald of a stranger.

THE ETERNAL THEME

Will he bring me business
Or will he bring me gladness
Or will he come for cure
Of his own sickness?

With a pedlar's burden
Walking up the garden
Will he come to beg
Or will he come to bargain?

Will he come to pester,
To cringe or to bluster,
A promise in his palm
Or a gun in his holster?

Will his name be John
Or will his name be Jonah
Crying to repent
On the island of Iona?

Will his name be Jason
Looking for a seaman
Or a mad crusader
Without rhyme or reason?

What will be his message—
War or work or marriage?
News as new as dawn
Or an old adage?

Will he give a champion
Answer to my question
Or will his words be dark
And his ways evasion?

Will his name be love
And all his talk be crazy
Or will his name be Death
And his message easy?

LOUIS MACNEICE

The Chariot

Because I could not stop for Death,
He kindly stopped for me;
The carriage held but just ourselves
And Immortality.

We slowly drove, he knew no haste,
And I had put away
My labour, and my leisure too,
For his civility.

We passed the school where children played,
Their lessons scarcely done;
We passed the fields of gazing grain,
We passed the setting sun.

We paused before a house that seemed
A swelling on the ground;
The roof was scarcely visible,
The cornice but a mound.

Since then 'tis centuries; but each
Feels shorter than the day
I first surmised the horses' heads
Were toward eternity.

EMILY DICKINSON

Ah! Sun-flower

Ah, Sun-flower! weary of time,
Who countest the steps of the sun;
Seeking after that sweet golden clime,
Where the traveller's journey is done;

Where the Youth pined away with desire,
And the pale Virgin shrouded in snow,
Arise from their graves, and aspire
Where my Sun-flower wishes to go.

WILLIAM BLAKE

Peace

My soul, there is a country
  Far beyond the stars,
Where stands a wingèd sentry
  All skilful in the wars:
There, above noise and danger,
  Sweet Peace sits crown'd with smiles,
And One born in a manger
  Commands the beauteous files.
He is thy gracious Friend,
  And—O my soul, awake!—
Did in pure love descend
  To die here for thy sake.
If thou canst get but thither,
  There grows the flower of Peace,
The Rose that cannot wither,
  Thy fortress, and thy ease.
Leave then thy foolish ranges;
  For none can thee secure
But One who never changes—
  Thy God, thy life, thy cure.

HENRY VAUGHAN

And Death Shall Have No Dominion

And death shall have no dominion.
Dead men naked they shall be one
With the man in the wind and the west moon;
When their bones are picked clean and the clean bones gone,
They shall have stars at elbow and foot;
Though they go mad they shall be sane,
Though they sink through the earth they shall rise again;
Though lovers be lost, love shall not;
And death shall have no dominion.

And death shall have no dominion.
Under the windings of the sea
They lying long shall not die windily;
Twisting on racks when sinews give way,
Strapped to a wheel, yet they shall not break;
Faith in their hands shall snap in two,
And the unicorn evil runs them through;
Split all ends up they shall not crack;
And death shall have no dominion.

And death shall have no dominion.
No more may gulls cry at their ears
Or waves break loud on the seashores;
Where blew a flower may a flower no more
Lift its head to the blows of the rain;
Though they be mad and as dead as nails
Heads of the characters hammer through daisies;
Break in the sun till the sun breaks down,
And death shall have no dominion.

DYLAN THOMAS

Fidele

Fear no more the heat o' the sun,
  Nor the furious winter's rages;
Thou thy worldly task hast done,
  Home art gone, and ta'en thy wages:
Golden lads and girls all must,
As chimney-sweepers, come to dust.

Fear no more the frown o' the great,
  Thou art past the tyrant's stroke;
Care no more to clothe and eat;
  To thee the reed is as the oak:
The sceptre, learning, physic, must
All follow this, and come to dust.

Fear no more the lightning-flash,
  Nor the all-dreaded thunder-stone;
Fear not slander, censure rash;
  Thou hast finish'd joy and moan:
All lovers young, all lovers must
Consign to thee, and come to dust.

No exorciser harm thee!
Nor no witchcraft charm thee!
Ghost unlaid forbear thee!
Nothing ill come near thee!
Quiet consummation have;
And renownèd be thy grave!

WILLIAM SHAKESPEARE

INDEX OF AUTHORS

INDEX OF AUTHORS

INDEX OF AUTHORS

INDEX OF FIRST LINES

INDEX OF FIRST LINES

INDEX OF FIRST LINES